Fabian Ruf

Excitonic Effects and Bandgap Instabilities in Perovskite Solar Cells

Excitonic Effects and Bandgap Instabilities in Perovskite Solar Cells

by
Fabian Ruf

KIT Scientific Publishing

Karlsruher Institut für Technologie
Institut für Angewandte Physik

Excitonic Effects and Bandgap Instabilities in Perovskite Solar Cells

Zur Erlangung des akademischen Grades eines Doktors der Naturwissenschaften von der KIT-Fakultät für Physik des Karlsruher Instituts für Technologie (KIT) genehmigte Dissertation

von M.Sc. Fabian Christopher Ruf

Tag der mündlichen Prüfung: 19. Juli 2019
Referent: Prof. Dr. H. Kalt
Korreferent: Priv.-Doz. Dr. M. Hetterich

This work has been supported financially by the Karlsruhe School of Optics and Photonics (KSOP) and the German Ministry of Education and Research (BMBF, FKZ 03SF0516C).

KSOP
Karlsruhe School of Optics & Photonics
http://www.ksop.de

Federal Ministry of Education and Research
http://www.bmbf.de
http://www.cisovskit.de

Impressum

KIT Scientific Publishing

Karlsruher Institut für Technologie (KIT)
KIT Scientific Publishing
Straße am Forum 2
D-76131 Karlsruhe

KIT Scientific Publishing is a registered trademark of Karlsruhe Institute of Technology.
Reprint using the book cover is not allowed.

www.ksp.kit.edu

Print on Demand 2020 – Gedruckt auf FSC-zertifiziertem Papier

ISBN 978-3-7315-1016-1
DOI 10.5445/KSP/1000105791

Excitonic Effects and Bandgap Instabilities in Perovskite Solar Cells

Zur Erlangung des akademischen Grades eines
DOKTORS DER NATURWISSENSCHAFTEN

von der KIT-Fakultät für Physik
des Karlsruher Instituts für Technologie (KIT)

genehmigte

DISSERTATION

von

M.Sc. Fabian Christopher Ruf
aus Karlsruhe

Tag der mündlichen Prüfung: 19.07.2019
Referent: Prof. Dr. H. Kalt
Korreferent: Priv.-Doz. Dr. M. Hetterich

Contents

1 Introduction **1**

2 Perovskite Absorber Materials for Thin-Film Solar Cells **5**
2.1 Organic–Inorganic Perovskites 5
2.1.1 Crystal Structure 6
2.1.2 Band Structure and Defects 8
2.1.3 Bandgap Tunability and Mixed Compounds 11
2.2 Fundamentals of Perovskite Solar Cells 13
2.2.1 Basic Structure and Working Principle 13
2.2.2 Cell Architectures and Fabrication Methods 16
2.2.3 Dynamic Hysteresis in j–V Analysis and Performance Characteristics of Perovskite Solar Cells 17
2.3 Opportunities and Challenges for Organic–Inorganic Perovskites 19
2.4 Summary and Relevance for This Work 22

3 Implications of Excitons on Dielectric Function and Absorption in Semiconductors **23**
3.1 The Dielectric Function of Semiconductors 23
3.2 Fundamentals of Excitons 26
3.3 Absorption of Excitonic Semiconductors and Elliott Formula 28
3.3.1 Bound States of Electrons and Holes 28
3.3.2 Ionization Continuum Above the Bandgap Energy 29
3.3.3 Elliott Formula and Sommerfeld Enhancement Factor 30
3.4 Summary 31

4 Absorption Spectroscopy and Evaluation Methods **33**
4.1 Experimental Realization of Temperature-Dependent Absorption Spectroscopy 33
4.2 Generalized Elliott Formula for the Evaluation of Absorption Spectra 35
4.3 f-Sum Rule Method for Determination of the Exciton Binding Energy 37
4.4 Summary 39

5 Modulation Spectroscopy of Semiconductors: Free Charge Carriers and Excitonic Effects . . . 41
5.1 Basic Principle of Modulation Spectroscopy . . . 41
5.2 Influence of Electric Fields on the Dielectric Function of Semiconductors . . . 44
5.2.1 Franz–Keldysh Theory . . . 44
5.2.2 The Low-Field Regime . . . 47
5.2.3 Confined States and Excitons . . . 49
5.3 Evaluation and Line Shapes . . . 50
5.3.1 Third-Derivative Functional Form (TDFF) . . . 50
5.3.2 First-Derivative Functional Form (FDFF) . . . 52
5.3.3 Modulus Transformation based on Kramers–Kronig Relations . . . 53
5.4 Summary and Comparison with Absorption Spectroscopy . . . 54

6 Electromodulation Spectroscopy of Perovskite Solar Cells . . . 57
6.1 Experimental Realization of Electromodulation Spectroscopy . . . 58
6.2 Influence of Measurement Geometry on Electromodulation Spectra . . . 61
6.2.1 Evaluation of Interference-Related Line-Shape Distortions in Electroreflectance Spectra . . . 62
6.2.2 Reflection and Transmission Configuration . . . 63
6.3 Identification of Suitable Modulation Voltages . . . 65
6.4 Detection at Fundamental and Second-Harmonic Frequency . . . 66
6.5 Summary . . . 71

7 Excitonic Nature of Optical Transitions in Perovskites . . . 73
7.1 Excitonic Origin of Fundamental Transitions in Electroabsoption Spectra of Perovskite Solar Cells . . . 74
7.1.1 Correlation of Absorption and Electroabsorption Spectroscopy . . . 74
7.1.2 Excitonic Line Shapes of Electroabsorption Spectra in Perovskites . . . 78
7.2 Exciton Binding Energy in Organic–Inorganic Perovskites . . . 81
7.2.1 Review of Published Values for the Exciton Binding Energy in Organic–Inorganic Perovskites . . . 82
7.2.2 Variation of Iodine–Bromine Ratio Used for Bandgap Tuning and its Impact on the Exciton Binding Energy . . . 83
7.2.3 The Exciton Binding Energy of Cesium-Containing Multiple-Cation Compounds . . . 86
7.3 Implications of Excitonic Effects for Solar-Cell Performance . . . 89
7.3.1 Increased Absorption Coefficient due to Sommerfeld Enhancement . . . 90
7.3.2 The Influence of Excitonic Effects on Charge-Carrier Extraction in Perovskite Solar Cells . . . 92
7.4 Summary . . . 95

8 Crystal-Phase Transitions in Organic–Inorganic Perovskites **97**
8.1 Phase Transitions in Methylammonium Lead Iodide and Methylammonium Lead Bromide 98
8.2 Phase-Transition Suppression in Cesium-Containing Mixed-Cation Compounds 101
8.3 Structural Verification of Phase-Transitions by Temperature-Dependent X-Ray Diffraction 103
8.4 Summary 106

9 Bandgap Instabilities in Multiple-Cation Mixed-Halide Perovskite Solar Cells **107**
9.1 Review of Studies on Compositional Stability of Mixed Perovskites 107
9.2 Reversible Shifts of the Bandgap Energy in Electroreflectance Spectra Under Illumination and Bias 109
9.2.1 Fundamental Phenomena 109
9.2.2 Influence of Different Absorber Compositions 111
9.3 Characterization of Photo- and Voltage-Induced Structural Changes 114
9.3.1 Correlation of Bandgap-Energy Shifts to Structural Changes Using X-Ray Diffraction 114
9.3.2 Time-of-Flight Secondary Ion Mass Spectrometry (ToF-SIMS) 117
9.4 The Role of Illumination and Voltage for Induced Bandgap Instabilities 119
9.4.1 Reversibility of Bandgap-Energy Shifts 120
9.4.2 Impact of Light and Applied Bias on Bandgap Instabilities 120
9.4.3 Effects Under DC Bias Conditions 122
9.5 Influence of Atmospheric Conditions on Bandgap Instabilities 123
9.5.1 Changes in Electroreflectance Spectra Under High Relative Humidity Conditions 123
9.5.2 Impact of Oxygen Content and Relative Humidity on Bandgap-Energy Shifts 125
9.6 Summary 127

10 Summary and Outlook **129**

A Electro-Optic Characterization Techniques and Experimental Details **133**
A.1 Verification of Temperature-Dependent Absorption Spectra by Means of a Commercial System with Integrating Sphere 133
A.2 Measurement and Evaluation of Temperature-Dependent Photocurrent Spectroscopy 134
A.3 Optical and Electronic Components of the Experimental Setups 137

B Reliability of Elliott Fits and *f*-Sum Evaluation Method **141**

C Analysis of Electromodulation Spectra Using First-Derivative Functional Form Line-Shape Fits **143**
C.1 Mathematical Expressions of Lorentzian- or Gaussian-Broadened FDFF Line Shapes 143
C.2 Comparison of Different Contributions in FDFF Line-Shape Fits of Experimental Electromodulation Spectra 145

D Numerical Simulation of Electroreflectance Spectra **149**

E Structural Characterization Methods and Experimental Details **155**
E.1 X-Ray Diffraction Analysis and Experimental Instrumentation 155
E.2 Time-of-Flight Secondary Ion Mass Spectrometry (ToF-SIMS) 156

F Samples Investigated In This Work **157**

List of Publications **165**

Bibliography **169**

Closing Words **211**

One shouldn't work on semiconductors,
that is a filthy mess; who knows whether
any semiconductors exist.

Wolfgang Pauli

The pace of research,
the pace of scientific and
technological advancement
with these [perovskite] materials
is like science on steroids.

Henry J. Snaith

If we knew what it was we were doing,
it would not be called research, would it?

Albert Einstein

Chapter 1

Introduction

The sun continuously irradiates the earth's surface with an energy flux of about 2×10^{17} Joule per second[1,2] – enough to cover the world's annual energy demand of about 200,000 TWh in just one hour.[3] This enormous energy source is completely sustainable and clean without any side effects such as the emission of greenhouse gases. In contrast, more than 85% of the world's primary energy consumption and 65% of worldwide electric-power generation are still fueled by the exhaustible fossil energy sources oil, natural gas, and coal.[3] The known reserves of these carbon-based fuels are estimated to be depleted within the next one hundred years based on the current annual consumption.[3,4] These numbers vividly convey an idea why a transformation of mankind's energy supply towards sustainable energy sources is inevitable in the coming decades. Moreover, they impressively illustrate the tremendous potential of applications harvesting solar energy. In particular with respect to electrical power generation, the direct conversion of light into electricity in photovoltaics[2,5,6] will play an increasingly important role in this transition process.[7,8]

Nowadays, crystalline silicon solar cells undisputedly dominate the market of photovoltaic technologies with a 95% share in production.[9] This success is based on power-conversion efficiencies of up to 26.7% for research cells and 24.4% for modules[9,10] and a comparably cheap production. However, the indirect bandgap of silicon is clearly not ideal and entails specific requirements for solar cells, primarily high-purity absorber layers with large thicknesses of several hundred micrometers to ensure sufficient light absorption.[6,11,12] These limitations led to the development of thin-film solar cells based on direct-bandgap semiconductors with thicknesses of only a few micrometers which open up novel applications and fabrication processes.[6,13] The most prominent representatives $Cu(In,Ga)Se_2$ and CdTe exhibit competitive power-conversion efficiencies up to 23.3%[14] and 22.1%,[10] respectively.

But recently, so-called *perovskite solar cells* entered the race. The term *perovskite* denotes a large material class with specific crystal structure including ferroelectric ceramics and high-temperature superconductors.[15–19] In the context of photovoltaic applications, metal halide perovskites, often also called organic–inorganic perovskites, are used which mostly consist of an organic cation within an inorganic metal–halide framework, e.g., methylammonium lead iodide as the most prominent representative.[20–22] In principle, these compounds are not new but were already synthesized and characterized, e.g., by D. Weber at the *University of Stuttgart* in 1978.[23] Still, it took several decades until their first application in photovoltaics.

Just a few months ago, perovskite-based solar cells "celebrated their 10th birthday": The very first journal report of a perovskite solar cell with 3.8% power-conversion efficiency was published by T. Miyasaka and co-workers on 14th of April in 2009.[24] They utilized a dye-sensitized solar-cell architecture and implemented methylammonium lead iodide as absorber dye because of its outstanding light absorption properties. Unfortunately, this first perovskite solar cell suffered from fast degradation due to the corrosive liquid electrolyte used. Its replacement with a solid-state hole-transporting material resulted in a greatly improved stability and a power-conversion efficiency of 10% bringing serious applications within reach for the first time.[25,26] Furthermore, the discovery of excellent ambipolar transport properties of both electrons and holes represents another important milestone[26,27] which eliminated the necessity of a mesoporous TiO_2 scaffold and led to new device architectures with a planar perovskite absorber layer.[19,28] This enabled a rather simple device fabrication based on spin-coating and subsequent low-temperature annealing.[29–31] As a result, perovskite solar cells attracted a great deal of attention and numerous research activities followed. This "perovskite fever" is, e.g., reflected by an exponentially increasing number of related published articles exceeding 3000 publications in 2018.[32] As a consequence, outstanding results have been achieved despite the comparable young age of this technology, e.g., a record research-cell efficiency of 24.2% outperforming $Cu(In,Ga)Se_2$ and CdTe.[10] Moreover, various other applications based on the excellent optoelectronic properties of organic–inorganic perovskites have been developed including light-emitting diodes (LEDs),[33–37] lasers,[38–45] and photodectors.[46–48] Or, to quote Henry J. Snaith, one of the most influential pioneers in perovskite photovoltaics: "The pace of research, the pace of scientific and technological advancement with these materials is like science on steroids."[49]

Nevertheless, because of the rapid growing-up of perovskite solar cells, a comprehensive understanding is still missing in many areas which will be essential to reach full maturity. This applies, for instance, to the exact solar-cell working mechanism. In this context, excitons, i.e., bound electron–hole pairs, play an important role since the separation and collection of charge-carriers is crucial for efficient solar cells. However, their binding energy and exact role is still heavily under debate. Another still existing key challenge – despite their exceptional success story – is the long-term stability of perovskite-based devices.[50–53] Although huge progress has been made increasing the stability to more than one year[54] and passing standard industry-relevant tests,[55–58] there is still a long way to go. Deeper insights into degradation mechanisms and related processes deteriorating solar-cell performance will surely further advance the perovskite technology.

In order to support this development, the present work focuses on a detailed electro-optic characterization of different perovskite compounds using optical spectroscopy.

Electromodulation spectroscopy is a powerful tool for precise investigations of the electronic structure of semiconductors. Developed about sixty years ago, its potential has been proven in various investigations of fundamental properties of semiconductors[59–66] or quantum-well structures.[67–73] In recent years, electromodulation spectroscopy was successfully applied even to multi-layer thin-film solar cells.[74–76] The possibility to study optical resonances in complete working devices can help to understand fundamental processes occuring in perovskite solar cells. In addition, in-depth evaluations of *absorption spectra* can further contribute to these explorations. Many experimental absorption spectra of metal halide perovskites surprisingly look like textbook examples of conventional high-quality semiconductors despite the comparably simple low-temperature solution-based fabrication.

Therefore, the objective of this thesis is to establish reliable measurements of absorption spectra and, in particular, electromodulation spectroscopy of complete perovskite solar cells. In a second step, both techniques will be combined to further reveal the role of excitons in organic–inorganic perovskites and implications for solar-cell performance. Moreover, structural changes in a semiconductor can affect its bandgap energy enabling the detection of the underlying processes using optical spectroscopy. Accordingly, absorption and electromodulation spectroscopy will be utilized to thoroughly examine temperature-induced crystal-phase transitions and reversible instabilities of the perovskite absorber in mixed compounds.

Structure of This Work

This thesis is divided into ten chapters. This general introduction is followed by a brief overview of organic–inorganic perovskites in chapter 2 describing their crystal and band structure as well as advantageous properties like defect tolerance and bandgap tunability. Furthermore, their implementation in solar cells is presented including basic working principle, fabrication methods, and performance aspects. Applications beyond solar cells and persisting challenges complement this chapter.

The foundation of both absorption and electromodulation spectroscopy is the interaction of light and matter. This can be modeled by the dielectric function which is introduced with a particular focus on excitonic implications in chapter 3.

Based on this, chapters 4 and 5 introduce experimental realization and evaluation methods of absorption and electromodulation spectroscopy. The analysis of absorption spectra includes a generalized version of the *Elliott formula* applicable for least-squares fitting of absorption spectra and an alternative so-called f-*sum rule* evaluation method. The Elliott formula in general describes the absorption coefficient of excitonic semiconductors.[77–79]

In case of modulation spectroscopy, a short general introduction is followed by a more detailed presentation of the influence of electric fields on the dielectric function since an external electric field (in the form of an applied bias) is used for electromodulation spectroscopy in case of perovskite solar cells. Again, the implications of excitons on expected experimental line shapes in the evaluation are highlighted.

The application of electromodulation spectroscopy to perovskite solar cells and important pecularities which have to be considered are discussed in chapter 6. This includes especially the detection at the second harmonic ($2f$) of the applied modulation frequency f.

In chapter 7, both temperature-dependent absorption and electromodulation spectroscopy are combined to reveal the influence of excitons on fundamental optical transitions. These rather proof-of-principle examinations of methylammonium lead iodide are followed by a quantitative comparison of the exciton binding energy in different mixed perovskite compounds utilizing the Elliott analysis introduced in chapters 3 and 4. This chapter is concluded by a discussion of beneficial and detrimental implications of excitons on the solar-cell performance including absorption and photocurrent spectra.

Temperature-dependent crystal-phase transitions can affect the electronic structure of semiconductors, e.g., the bandgap energy. This is exploited for optical detection of phase transitions in different perovskite compounds using absorption and electromodulation spectroscopy in chapter 8 and confirmed by X-ray diffraction results.

Chapter 9 presents a comprehensive study of a reversible decrease of the bandgap energy discovered in multiple-cation mixed-halide perovskite compounds. A reversible segregation of the perovskite into iodine-rich and bromine-rich domains under 1 sun illumination and applied voltage is identified as origin of the observed bandgap instabilities using electromodulation spectroscopy, X-ray diffraction and secondary ion mass spectrometry. A detailed characterization of the driving forces and atmospheric influences, i.e. relative humidity and oxygen content, and a discussion of implications for solar-cell performance complement these investigations.

In the final chapter, this thesis is concluded by summarizing the main results and indicating possible follow-up investigations.

Chapter 2

Perovskite Absorber Materials for Thin-Film Solar Cells

With the very broad area of research and application of organic–inorganic perovskites in mind, this chapter aims at giving a brief overview about structural and electronic characteristics of this material class including crystal phases and band structure (sections 2.1.1 and 2.1.2). The extended field of use is based on the tunability of its composition and related properties, foremost the bandgap energy (section 2.1.3). Organic–inorganic perovskites have attracted a great deal of attention in the wake of their integration into a solid-state solar cell leading to a 10% power-conversion efficiency in 2012.[25,26] Accordingly, they are predominantly analyzed in the context of photovoltaic applications within this work. The working principle of perovskite solar cells mainly relies on charge-selective layers and can be described by a p–i–n heterojunction (section 2.2.1). However, mobile ions in the device lead to a more complex situation because they can influence electronic transport and device characterization (section 2.2.3). Furthermore, commonly used solar-cell architectures and possible fabrication processes are briefly presented (section 2.2.2). The chapter is complemented by highlighting further applications of organic–inorganic perovskites also beyond photovoltaics based on their advantageous characteristics, primarily exceptional optoelectronic properties and versatile fabrication methods. However, perovskite-based devices still face open challenges such as toxicity and stability which are also presented (section 2.3).

2.1 Organic–Inorganic Perovskites

The name perovskite refers to the crystal structure of calcium titanate ($CaTiO_3$) and denotes all compounds with the same type of crystal structure.[15–18] In general, this includes, i.a., ferroelectric ceramics such as $BaTiO_3$ as well as high-temperature superconductors.[17–19] In the context of solar cells and related optoelectronic devices, so-called organometal or (hybrid) organic–inorganic halide perovskites incorporating organic cations as well as inorganic metal and halide ions are the relevant class of perovskite materials (Solar cells based on this type of perovskites as absorber layer are often shortly denoted as "perovskite solar cells".). The most prominent representative is by far methylammonium lead iodide which is discussed as a model system in the following.

2.1.1 Crystal Structure

Perovskites are described by the formula $\mathbf{ABX_3}$. In case of organic–inorganic compounds, **A** denotes an organic cation, **B** a metal cation (mainly Pb^{2+} or Sn^{2+}), and **X** a halide anion (I^-, Br^-, Cl^-) – leading to $CH_3NH_3PbI_3$ (or abbreviated $MAPbI_3$ with $MA^+ \mathrel{\widehat{=}} CH_3NH_3^+$) as model system methylammonium lead iodide.[20–23] Figure 2.1 visualizes the cubic crystal structure with organic MA^+ cation in the center of the unit cell and one exemplary PbI_6 octahedron (marked in blue). Other possible substitutes for **A**, **B** and **X** are discussed in section 2.1.3. The simple cubic structure is only observed at high temperatures above 327 K[80] at which the spherically non-symmetric MA^+ cation oscillates between different rotational orientations on a picosecond time scale resulting in, on average, spherical symmetry.[81–83] At lower temperatures, the reduced rotational motion of the MA^+ cation leads to distortions of the inorganic Pb–I cage and, thereby, a tetragonal or orthorhombic structure with reduced symmetry.

Below 162 K, the MA^+ cations are in a fully ordered state due to stabilization by hydrogen bonding between the NH_3 endgroup and the iodide of the inorganic cage.[82, 84, 85] This alignment of the MA^+ cations causes a distortion of the inorganic Pb–I framework because of the missing rotational symmetry of MA^+ ions (see I–Pb bonding angle in Fig. 2.2 (a) left) which is accompanied by a tilting of the PbI_6 octahedra (see right-hand side of Fig. 2.2 (a)). As a result, this leads to an orthorhombic phase with reduced symmetry compared to the cubic structure.[82, 84–86] Since the MA^+ cations of two layers on top of each other are aligned antiparallel (see Fig. 2.2 (a)), a larger unit cell compared to the simple cubic structure is necessary to describe the full structure.[85, 87, 88]

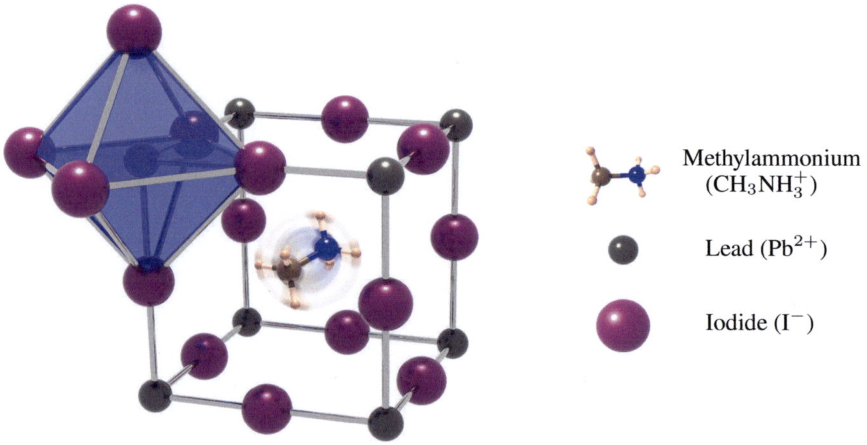

Figure 2.1: The cubic perovskite structure $\mathbf{ABX_3}$ with organic cation **A** in the center, here methylammonium ($CH_3NH_3^+$ with C in brown, N in blue, and H in beige), metal cation **B** (gray), here lead (Pb^{2+}), and halide anion **X** (purple), here iodide (I^-). An exemplary PbI_6 octahedron is marked in blue. Rotations of the organic cation on a picosecond time scale at high temperatures above 327 K result in, on average, spherical symmetry ensuring a cubic structure (as shown here).

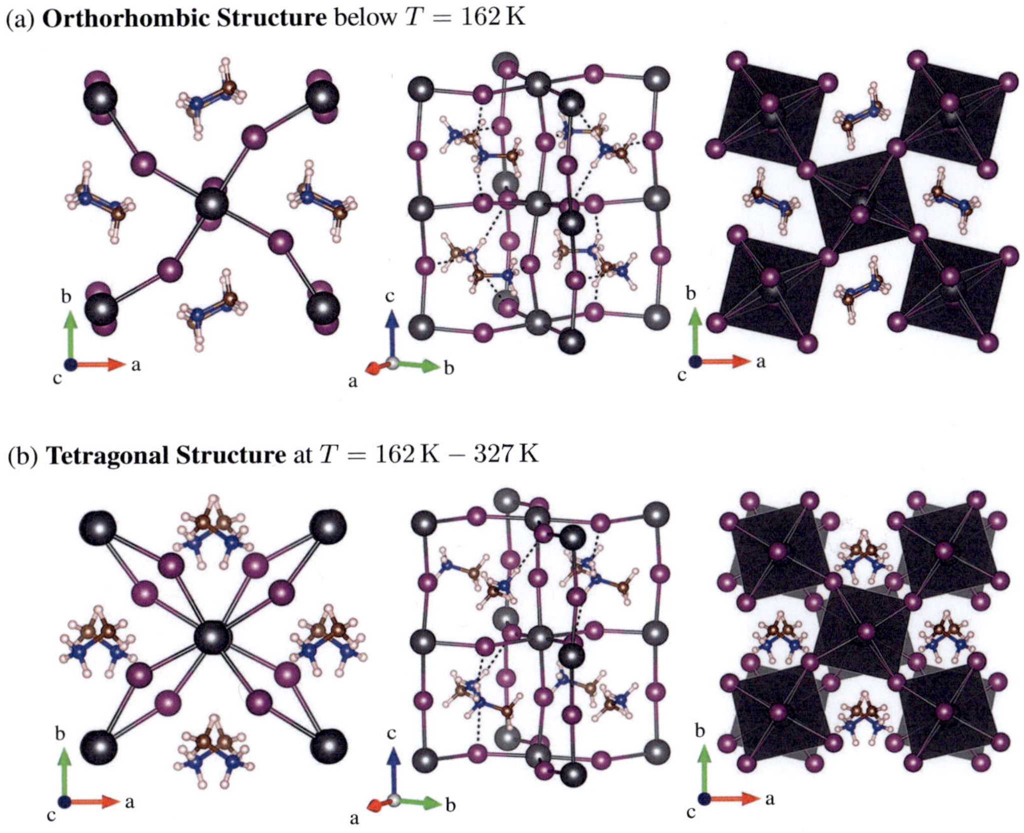

Figure 2.2: Top view, side view, and polyhedron graphs of (a) low-temperature orthorhombic and (b) room-temperature tetragonal crystal structure of $MAPbI_3$ with Pb^{2+} (gray), I^- (purple) and $CH_3NH_3^+$ (C in brown, N in blue and H in beige). Adapted with permission from [87]. Copyright 2014 American Chemical Society.

Between 162 K and 327 K, $MAPbI_3$ crystallizes in a tetragonal structure (illustrated in Fig. 2.2 (b)). The increased thermal motion of the MA^+ cation weakens the hydrogen bonds and results in more possible orientations of MA^+ than in the orthorhombic phase.[82,84] As a result, the Pb–I framework is less distorted which can be recognized by comparing the tilting of the PbI_6 octahedra (especially along the c axis) in the tetragonal and orthorhombic structure (see right-hand side of Fig. 2.2 (a) and (b)). Still, the exact configuration of MA^+ is not unambiguously clear in literature.[82,84,85,87,89,90]

The implications of temperature-induced transitions between these crystal phases in $MAPbI_3$ and other perovskite compounds on the optoelectronic properties such as the bandgap energy are discussed in chapter 8.

2.1.2 Band Structure and Defects

The band structure of perovskites has been calculated by various research groups using *density functional theory* (DFT) at different levels of complexity.[87,88,91–100] Figure 2.3 presents results for the tetragonal crystal structure of $MAPbI_3$ at room temperature (left part).[94] The direct bandgap is at the Γ point in the center of the Brillouin zone[1].[81,94] The bandgap energy E_g of $1.5\,eV - 1.6\,eV$ fits well to experimental results (see also chapters 7 and 8).

The breakdown of the *density of states* (DOS) into contributions of different atomic orbitals (right part of Fig. 2.3) reveals one of the exceptional properties of lead halide perovskites:

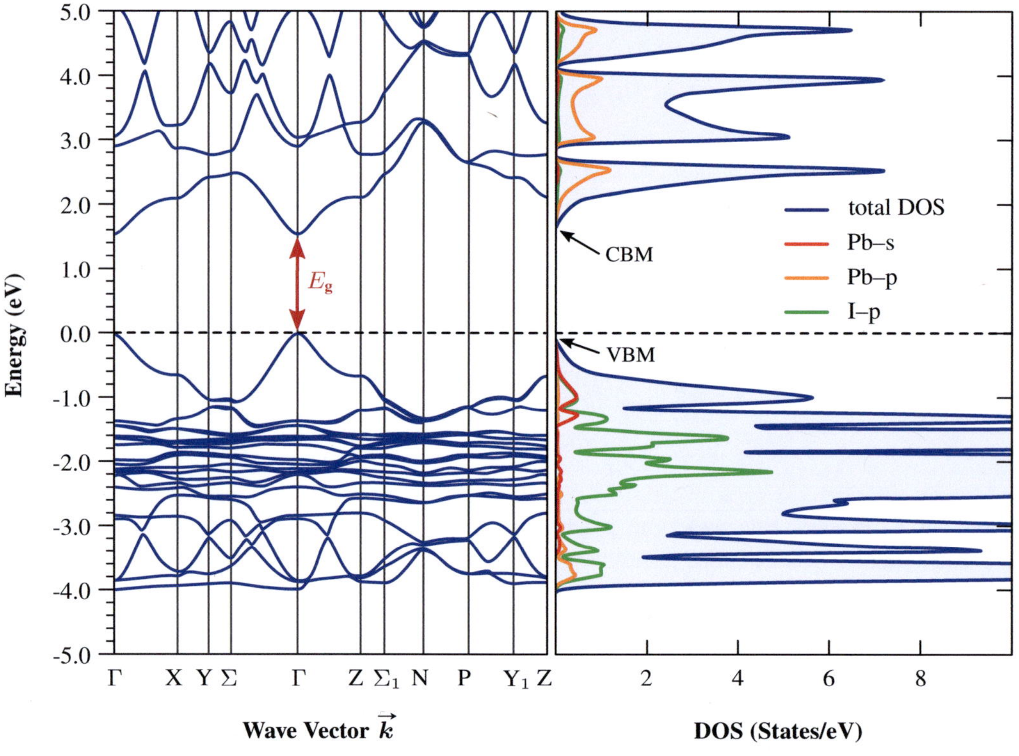

Figure 2.3: Calculated band structure and density of states (DOS) for the tetragonal phase of $MAPbI_3$. The bandgap is at the Γ point. The DOS is dominated by Pb–s and I–p orbitals at the valence band maximum (VBM) and Pb–p (and, to some extent, I–p) orbitals at the conduction band minimum (CBM). Adapted with permission from [94]. Copyright 2014 American Chemical Society.

[1] In the cubic structure, it is located at the R point which can be zone-folded back onto the Γ point of the tetragonal or orthorhombic structure.

The valence band maximum (VBM) is mainly composed of Pb–s and I–p orbitals, and the conduction band minimum (CBM) is formed by Pb–p (and, to some extent, I–p) orbitals. The p orbitals forming the CBM show less dispersion compared to s orbitals in the CBM of, e.g, gallium arsenide (GaAs). The resulting larger DOS in the CBM (leading to a higher joint density of states, see chapter 3.1) is one of the reasons for the exceptional absorption of perovskites.[93, 101] Additionally, the intra-atomic transitions from Pb–s to Pb–p states possess high transition probabilities.[101] The impact of the organic MA^+ cation on the electronic structure is negligible except from indirect effects due to distortion of the Pb–I inorganic framework as discussed above.[81, 87, 93, 102]

The bonding diagram in Fig. 2.4 (a) reveals the nature of the molecular orbitals (blue) forming VBM and CBM from the Pb–6s and I–5p atomic orbitals (gray) and the Pb–6p and I–5p orbitals (gray), respectively: Both are *antibonding*[91, 103–105] – in contrast to other typical semiconductors such as CdSe or GaAs (see Fig. 2.4 (b)).[106]

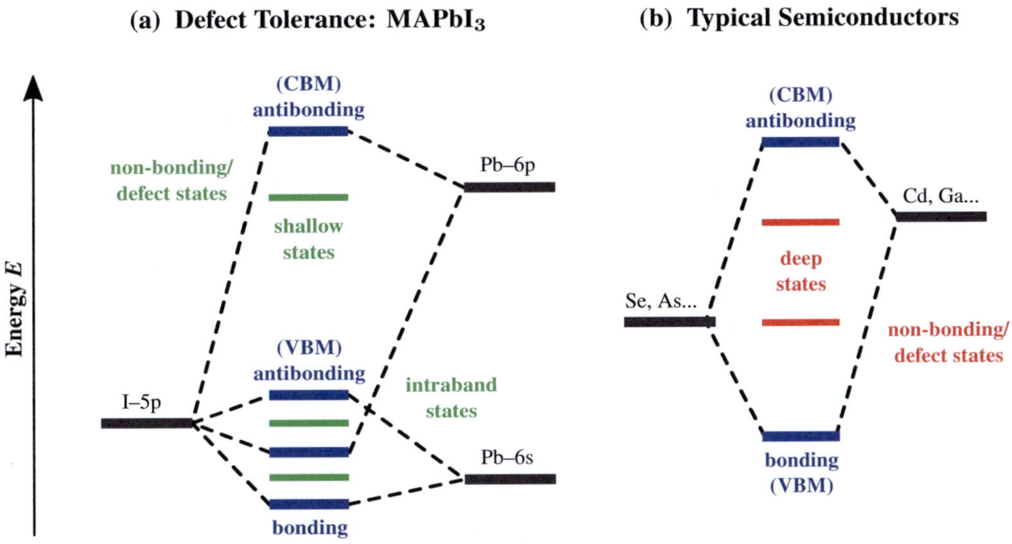

Figure 2.4: (a) Bonding diagram of $MAPbI_3$ illustrating the molecular orbitals (blue) composed of I–5p, Pb–6s, and Pb–6p atomic orbitals (gray). Both CBM and VBM are formed by antibonding molecular orbitals of lead and iodine. As a result, vacancies and corresponding weakly or non-bonding orbitals (corresponding energy levels in green) only lead to shallow or intraband defect states which is known as the *defect tolerance* of lead halide perovskites. (b) For comparison, in typical semiconductors such as CdSe or GaAs, the VBM is formed by bonding molecular orbitals (here as simplified schematic diagram). Thereby, weakly or non-bonding orbitals due to vacancies result in defect states (red) within the bandgap. See also [105, 106].

This is the origin of the so-called *defect tolerance* of lead halide perovskites which essentially means that defects do not strongly affect the electrooptic properties.[104–107] In case of typical semiconductors such as CdSe or GaAs, vacancies (e.g., Cd vacancies) result in defect states (red) within the bandgap due to remaining weakly or non-bonding orbitals (e.g., of Se) as illustrated in Fig. 2.4 (b).[104–106] These deep traps then act as charge-carrier recombination centers, e.g., for Shockley–Read–Hall recombination.[103] In contrast, the bandgap in lead halide perovskites is formed between two antibonding molecular orbitals and, therefore, these defect states (green) mainly reside inside the valence or conduction band or form only shallow traps close to the band edge.[104–106, 108] This is one of the reasons for the outstanding optoelectronic properties of lead halide perovskites despite the low-temperature solution-based fabrication processes (e.g., compared to high-purity silicon or GaAs). Furthermore, the antibonding nature of CBM and VBM has a strong impact on the temperature dependence of the bandgap energy E_g which will be discussed in chapter 8.1.

Nevertheless, there are defects present in halide perovskites which have been studied by numerous theoretical calculations. Antisite substitution of iodine on MA or lead sites and vice versa (I_{MA}, I_{Pb}, MA_I, Pb_I) are found to be energetically and kinetically unstable and are expected to dissociate into vacancies and interstitials.[109, 110] Accordingly, interstitials (I_i, MA_i, Pb_i) and vacancies (V_I, V_{MA}, V_{Pb}) are the most likely formed defects.[109, 110] The reported *defect formation energies* are not fully consistent which might be partly caused by the calculation itself but is also related to different conditions considered, e.g, non-stochiometric compounds.[111–114] For instance, some studies predict lower defect densities and less detrimental effects for slightly I-poor (and, therefore, Pb-rich) fabrication.[115, 116] Other investigations even report the possible formation of deep-level defects[116, 117] which could be inactivated under certain fabrication conditions or post-synthesis treatment.[116] Experimental examinations (based on admittance spectroscopy[118, 119] or thermally stimulated current measurements[120, 121]) yield defect activation energies between 70 meV and 700 meV with trap densities in the order of $10^{15}\,\mathrm{cm}^{-3} - 10^{17}\,\mathrm{cm}^{-3}$ (multicrystalline silicon: $10^{12}\,\mathrm{cm}^{-3} - 10^{15}\,\mathrm{cm}^{-3}$ [121–123]).

However, there is general consensus that iodine is the most mobile ion in $MAPbI_3$ followed by MA based on calculated activation energies.[22, 110, 113, 114, 124–126] This *ion migration* is easily activated and has also been observed experimentally.[127–129] Still, further investigations are needed to clearly identify the migration channels, e.g., in the bulk and/or – as suggested by recent results – at grain boundaries.[110, 124, 125, 130] The good conductivity for ions in halide perovskites also affects the charge-carrier dynamics in solar cells and, thereby, influences their working mechanism (see section 2.2.1). Moreover, mobile ions are most likely at least one of the reasons for the phenomena of hysteresis effects in j–V charactistics (see section 2.2.3) and reversible halide segregation in mixed compounds (see chapter 9). Furthermore, other ions initially not present in the perovskite itself have been found to migrate into it, e.g., Au atoms from the back contact in solar cells leading to severe degradation.[114, 131]

2.1.3 Bandgap Tunability and Mixed Compounds

Undoubtly, one of the major advantages of perovskites is the possibility to integrate different cations and anions in the $\mathrm{ABX_3}$ crystal structure which opens up a large space for tailor-made properties, e.g., the bandgap energy E_g.

The so-called *tolerance factor* t introduced by Goldschmidt[20, 132, 133] is based on a solid-sphere model and provides guidance in terms of formability and expected crystal structure of possible compounds:

$$t = \frac{R_\mathrm{A} + R_\mathrm{X}}{\sqrt{2}\,(R_\mathrm{B} + R_\mathrm{X})} \tag{2.1}$$

with effective radii R_A, R_B, and R_X of A, B, and X ion, respectively. For $0.81 < t < 1.1$, the formation of a perovskite compound can be expected.[20, 134] It presumably crystallizes in a cubic structure for the narrower range of $0.9 < t < 1.0$ or in the tetragonal phase for smaller t ($0.81 < t < 0.9$).[20, 135] For more precise predictions, an additional *octrahedral factor* $\mu = R_\mathrm{B}/R_\mathrm{X}$[133, 134, 136] can be utilized which needs to fulfill the condition $0.44 < \mu < 0.90$ for perovskite formation.[20, 134]

Figure 2.5 presents the "perovskite toolbox" of widely used cations and their effective radii.[20, 135, 137] While the variety of the usable B-site cations lead and tin (and, potentially, germanium[138]) and X-site anions iodine, bromine, and chlorine is rather limited (at least for optoelectronic applications), many more small organic molecules than the most commonly-used methylammonium (MA), formamidinium (FA), cesium, and rubidium are possible.[138]

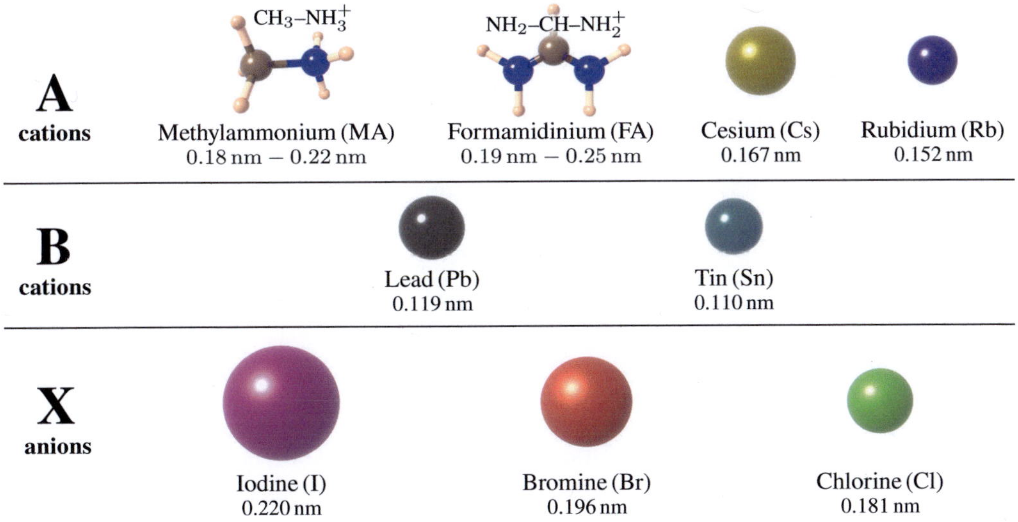

Figure 2.5: Commonly used "perovskite toolbox" of cations (**A**, **B**) and anions (**X**) for mixed perovskite compounds including effective ionic radii.[20, 135, 137]

Furthermore, numerous combinations of the different ions and ratios of mixed cations and/or anions are possible. In particular, some "pure" compounds containing only one type of ion at A, B, and X site, e.g., $FAPbI_3$ or $CsPbI_3$ do not form stable photoactive perovskite phases[86, 139, 140] but can be stabilized by incorporation of small ratios of other cations.[141–144]

One of the main benefits of composition tuning is the possibility of engineering the related bandgap energy towards desired values. For instance, the bandgap energy E_g of $MAPbI_3$ of about $1.6\,eV$ is still larger than the optimum value of $1.3\,eV - 1.5\,eV$ of the Shockley–Queisser limit.[145, 146] In this case, incorporation of the larger FA cation helps to decrease the bandgap energy.[143, 147] At the other end, organic–inorganic perovskites are ideal candidates for wide-bandgap top cells in tandem applications with, e.g., silicon or $Cu(In,Ga)Se_2$ bottom cells.[57, 148–154] In this configuration, slightly increased values of E_g of $1.65\,eV - 1.80\,eV$ are desirable[155, 156] which can be achieved, e.g., by substitution of iodine with bromine. By tuning the bromide content x in $MAPb(I_{1-x}Br_x)_3$, the bandgap energy can be adjusted in a wide range of $1.6\,eV - 2.3\,eV$[157–159] (see Fig. 7.6) which can be also interesting for optoelectronic applications beyond photovoltaics, e.g., perovskite-based lasers[38–41, 43–45] or light-emitting diodes (LEDs).[34–37, 160] By additional (partial) substitution of the B-site cation lead with tin, even perovskite–perovskite tandem cells with power-conversion efficiencies up to 23% have been realized (bottom cell: $E_g \approx 1.2\,eV - 1.3\,eV$, top cell: $E_g \approx 1.7\,eV - 1.8\,eV$).[161–166]

In addition to tailoring the bandgap energy for specific applications, perovskite compounds with mixed cations and/or anions have demonstrated reduced defect densities and improved charge-carrier transport[120] which leads to enhanced power-conversion efficiencies (PCE). The model example in this context is the incorporation of cesium and rubidium in the already known $(FA,MA)Pb(I,Br)_3$ mixed-compound leading to a PCE up to 21.6% and improved reproducibility.[167, 168] On top of that, the long-term stability of the devices is significantly increased to over 500 hours (preserving 95% of the initial PCE) under continuous illumination at 85 °C (compared to $< 100\,h$ without Cs and Rb).[167, 168] However, while there is broad consensus about the benefits of cesium incorporation,[167–171] the additional rubidium could lead to unwanted non-photoactive non-perovskite side phases since it is slightly to small for incorporation in a perovskite structure according to the tolerance factor.[172] Detailed studies of compositional stability of Cs-containing multiple-cation mixed-halide perovskites are presented in chapter 9.

In a similar fashion, various additional cations have been investigated – some are incorporated in the perovskite structure, others rather act as doping or lead to improved crystallization.[173–176] Further details can be found, e.g., in [138, 140, 177]. In particular, so-called *2D perovskites* have recently gained significant attention. By intentionally adding organic molecules which are too large to fit in a perovskite structure (e.g., $C_6H_5(CH_2)_2NH_3^+$), compounds consisting of alternating organic and inorganic layers are formed.[140, 178–182] The layered structure reduces moisture sensitivity and degradation but also decreases charge-carrier transport.[140, 178, 181, 183] The combination of conventional 3D and layered 2D perovskites promises both high PCE (up to 21% PCE) and enhanced long-term stability (95% of initial PCE after 500 hours of operation).[54, 183–187]

2.2 Fundamentals of Perovskite Solar Cells

Thin-film solar cells are most likely the most prominent application of organic–inorganic perovskites due to the outstanding rise of their power-conversion efficiency outperforming (on a reasearch scale) established technologies such as $Cu(In,Ga)Se_2$ and multicrystalline silicon. This chapter will briefly introduce the basic structure and working principle of perovskite-based solar cells and present possible fabrication methods, cell architectures and performance issues.

2.2.1 Basic Structure and Working Principle

In general, the basic principle of a photovoltaic device can be divided into [1, 188–190]

(i) generation of electron–hole pairs due to absorption of photons and
(ii) separation and extraction of electrons and holes at different contacts.

Reflecting about step (i) with regard to organic–inorganic perovskites, the outstanding absorption coefficient in the range of values for GaAs (or even above)[20, 191] and its relation to the specific nature of valence and conduction band have been already discussed in section 2.1.2. As a consequence, thicknesses of the perovskite absorber in solar cells as low as $300\,\mathrm{nm} - 400\,\mathrm{nm}$ are sufficient for devices with high power-conversion efficiency exceeding 22%.[30, 167, 168, 192–194] Additionally, comparable low exciton binding energies[76, 195] (e.g., compared to organic semiconductors[196, 197]) enable thermal dissociation of bound electron–hole pairs resulting in predominantly free electrons and holes under working conditions (room temperature, 1 sun illumination). See chapter 7 for a detailed discussion of excitonic effects. In summary, the efficient generation of electron–hole pairs and separation into free electrons and holes is easily possible in organic–inorganic perovskites and one of their major advantages.

In a second step (ii), the photogenerated electrons and holes have to be driven towards different contacts in order to convert their energy into usable electric power.[1] To achieve selective transport of electrons (holes) towards "their" respective electrode, an electron (hole) transport layer with high conductivity for electrons (holes) and low conductivity for holes (electrons) is required.[1, 189, 190, 198, 199] These layers can be perceived as "semipermeable membranes" essentially blocking either electrons or holes and can be realized in a number of ways, e.g., by doping in a p–n homojunction or a more effective p–i–n heterojunction, or in *dye-sensitized solar cells* (DSSC).[1, 189, 190, 199]

Historically, today's perovskite solar cells evolved from liquid-electrolyte DSSC with $MAPbI_3$ or $MAPbBr_3$ as light-absorbing dye on a mesoporous TiO_2 scaffold reaching power-conversion efficiencies between 0.4% and 3.8%.[19, 24, 200, 201] By substituting the liquid electrolyte with spiro-OMeTAD[2] as a solid-state hole-transport layer, power-conversion efficiency could be further increased to 10% with simultaneously enhanced stability due to the removal of the corrosive liquid electrolyte.[25, 26] Furthermore, Snaith and co-workers replaced the electron-transporting

[2] 2,2',7,7'-tetrakis-(N,N-di-p-methoxyphenyl-amine)-9,9'-spirobifluorene

TiO_2 with an insulating Al_2O_3 scaffold. Surprisingly, the power-conversion efficiency was even increased which impressively demonstrated that the perovskite material *itself* facilitates good ambipolar transport of both holes *and* electrons.[26,27] This is also reflected by large electron and hole mobilities and diffusion lengths exceeding $1\,\mu m$.[86,202–206] This inherent property of organic–inorganic perovskites led to the realization of solar-cell architectures with a planar perovskite absorber film[3] sandwiched between an electron- and a hole-extraction layer[19,28] which is schematically depicted in Fig. 2.6. Electron–hole pairs are photogenerated in the perovskite absorber and collected at the respective metal or TCO[4] contact. The *electron-transport layer* (ETL) and *hole-transport layer* (HTL) act as the required charge-selective layers due to their energy bands blocking either holes or electrons, respectively.

Although there is still an ongoing debate about the exact working principle[208–211] perovskite solar cells are often perceived as n–i–p heterojunctions with the perovskite absorber as a high-quality intrinsic layer.[212–214] Figure 2.7 (a) and (b) illustrate the band diagram of a n–i–p heterojunction in the dark and under illumination at the operating voltage for maximum power output, the so-called *maximum power point* (MPP), respectively. The n-doped and p-doped layer can be identified with ETL and HTL, respectively. Under equilibrium conditions in the dark (Fig. 2.7 (a)), the difference in the *work functions* of n-doped and p-doped layer leads to a *built-in potential* V_{bi}.[199,211,215] Under illumination (Fig. 2.7 (b)) the Fermi level splits into separate quasi-Fermi levels for electrons and holes $E_{F,n}$ and $E_{F,p}$, respectively, within the intrinsic layer which determine the maximum achievable voltage $V = (E_{F,n} - E_{F,p})/e$.[199,211,215] Small gradients in these quasi-Fermi levels act as driving force for the charge carriers, and selective collection of electrons and holes is achieved by the extraction layers.[1,199]

However, the situation appears to be more complex in case of perovskite solar cells since additional mechanisms can modify the well-known n–i–p working principle. Here, in particular, the high ionic conductivity in perovskites has to be considered which leads to the presence of mobile ions, predominantly iodine in $MAPbI_3$ (see section 2.1.2). These additional movable charges can influence internal electric fields and, e.g., lead to the complete screening of electric fields within the bulk of the perovskite layer by accumulation of mobile ions at the interfaces between perovskite and ETL/HTL as illustrated in Fig. 2.7 (c).[216,217] In general, the coupling of the electronic transport of holes and electrons with an additional comparably slow transport of mobile ions, can affect the charge-carrier extraction under working conditions. Moreover, the accumulation of ions at interfaces will influence surface recombination and, thereby, depend on the actual materials and interface quality.[217] Various simulations have been performed to elucidate these more complex phenomena in perovskite solar cells – including mobile cations and/or anions, different boundary conditions, and different doping of the perovskite (which might be unintentionally p- or n-doped).[112,213,217,219–222] Despite good agreement in most cases further studies are needed for an universal model beyond individual experimental and computational results.

[3] In some device structures, an additional mesoporous TiO_2 between compact TIO_2 and perovskite absorber (also infiltrated by the perovskite) is still used because it seems to improve the interface. However, it is not necessary for well-performing solar cells as pointed out above.[207,208]

[4] *Transparent Conductive Oxide*, e.g., FTO (**F**luorine-doped **T**in **O**xide), ITO (**I**ndium-doped **T**in **O**xide), or IZO (**I**ndium-doped **Z**inc **O**xide)

In summary, organic–inorganic perovskites exhibit outstanding inherent properties for solar cells, namely, high absorption coefficient, sufficiently low exciton binding energy and excellent ambipolar transport of both electrons and holes. In addition, suitable electron- and hole-transport layers are crucial for high-performance devices. Moreover, the fabrication of high-quality films and interfaces of both absorber and extraction layers is important to avoid detrimental effects such as non-radiative (surface) recombination.

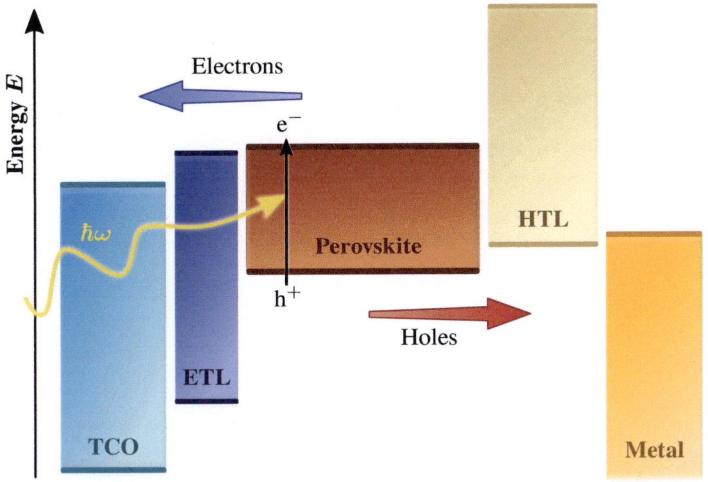

Figure 2.6: Basic structure of a perovskite solar cell. Incident light (from left side) generates electron–hole pairs in the perovskite absorber. ETL and HTL ensure selective transport of electrons and holes towards TCO and metal contact, respectively. See also [20, 218].

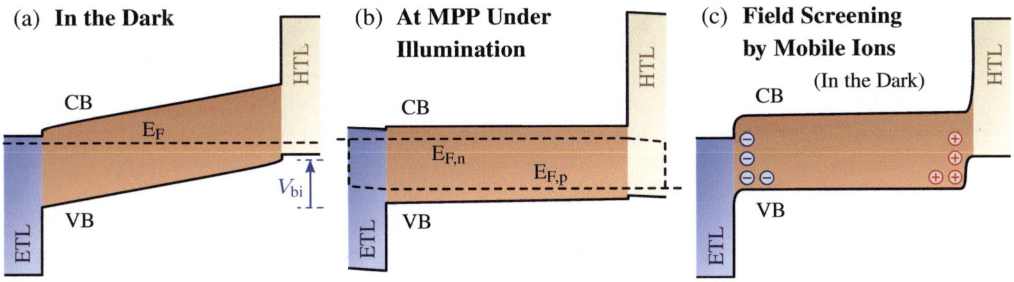

Figure 2.7: Illustration of a n–i–p heterojunction (here, ETL–perovskite–HTL) (a) in dark and (b) under illumination at maximum power point. Under illumination, the Fermi level splits into quasi-Fermi levels for electrons and holes $E_{F,n}$ and $E_{F,p}$ within the absorber layer. (c) Mobile ions present in perovskites presumably introduce different additional effects, e.g., screening of electric fields within the bulk of the perovskite layer due to accumulation of ions at the interfaces (in the dark). See also [199, 216].

2.2.2 Cell Architectures and Fabrication Methods

Perovskite solar cells can be fabricated in various configurations using different materials and layer sequence. The most commonly used so-called "standard" architecture is illustrated in Fig. 2.8 (a).[29–31] The layer stack is built up on a glass substrate coated with a transparent conductive oxide, mostly FTO[5] or ITO[6], as front contact. Typically used materials for the electron-selective layer (ETL) are TiO_2 and SnO_2, and spiro-OMeTAD is mainly used as hole-selective layer (HTL). The device is completed by a thin gold layer as back contact. The final solar cells are illuminated from the glass side and electrically contacted at the gold and FTO[5] (ITO[6]) layer. Figure 2.8 (b) shows an exemplary scanning-electron micrograph of the sample structure mainly investigated within this work.

Moreover, also semitransparent devices are achievable by substitution of the gold back contact with a TCO layer, e.g., sputtered IZO[7] [223] (see Fig. 2.8 (c)). Thereby, fabrication of tandem solar cells with perovskites as large-bandgap absorber can be realized.[148, 151, 153, 154] Furthermore, optical spectroscopy techniques in transmission configuration become possible which are applied within in this work (see chapter 7).

Standard Architecture **Semitransparent Cell**

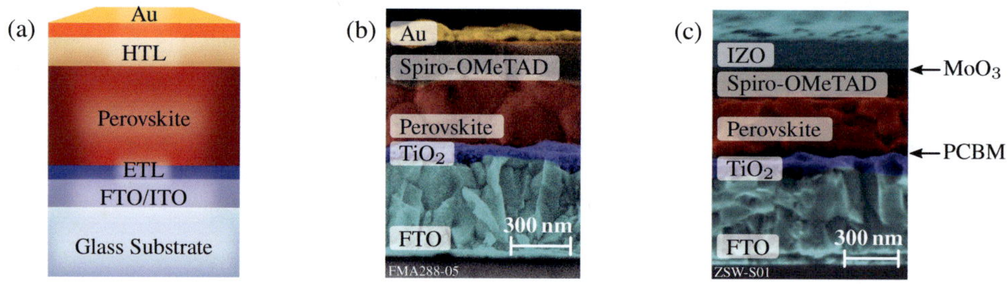

Figure 2.8: (a) Standard layer stack of a perovskite solar cell with ETL / perovskite / HTL / gold deposited on a glass substrate coated with FTO[5] or ITO[6]. (b) Scanning-electron micrograph of an exemplary device mainly used within this work. (c) Semitransparent devices by substitution of the gold contact with sputtered IZO allow optical spectroscopy in transmission as well as fabrication of tandem solar cells (Thin additional interlayers of MoO_3 between IZO[7] and spiro-OMeTAD, and PCBM[8] between TiO_2 and perovskite are implemented but are not visible here.). (b) and (c) have been obtained by Simon Woska and Eva Wirth at KIT, and Moritz Schultes at the *Zentrum für Sonnenenergie- und Wasserstoff-Forschung Baden-Württemberg* (ZSW) in Stuttgart, respectively.

---

[5] **F**luorine-doped **T**in **O**xide
[6] **I**ndium-doped **T**in **O**xide
[7] **I**ndium-doped **Z**inc **O**xide
[8] **P**henyl-**C**61-**B**utyric-acid-**M**ethyl ester, a fullerene derivative

Additionally to the standard architecture, an inverted structure with exchanged order of electron- and hole-selective layer is feasible which enables the use of other charge-carrier extraction layers inspired by organic solar cells.[31,224] Promising results in terms of performance and stability have been recently reported for solar cells with inverted architecture.[225–230]

Apart from absorber composition and layer sequence, enormous efforts have been invested in the investigation of improved charge-selective layers (predominantly HTLs)[231] in order to enhance stability, e.g., by inorganic layers such as NiO[228,232] or CuSCN,[233] increase performance,[213,234,235] and reduce costs.[236,237]

The vast majority of perovskite solar cells and, in particular, their absorber layer are fabricated by spin-coating of a solution with dissolved precursor salts and subsequent thermal annealing to evaporate residual solvents and start crystallization.[29–31] The straightforward and rather fast sample preparation facilitates low-cost production and still yields high-quality layers and devices. Various processing steps and parameters have been developed to improve crystal quality, e.g., one or two step deposition or solvent engineering.[29–31,238–240] All samples investigated within this work are based on solution-processing. See appendix F and [223,241] for more details about the layer stack. Additionally, solvent-free fabrication of perovskite absorber layers and complete devices is possible using vacuum deposition. Thereby, the mostly toxic solvents are avoided and a better control of layer thicknesses and composition as well as more flexibility in terms of layer sequence[242] is achieved.[28,243–246]

Particularly with regard to commercial applications, numerous studies examined up-scalable high-throughput fabrication methods. Spray coating,[247,248] slot dye coating,[249–251] and blade coating[252–254] present promising steps towards fast and cheap roll-to-roll processing. Additionally, printing techniques such as screen printing,[255] inkjet printing,[41,256,257] and aerosol-jet printing[258] offer the possibility of customizable and complex geometries. By additional inkjet-printed fluorescent dye layers, variously colored perovskite solar cells make further practical applications accessible,[259] e.g, for building-integrated photovoltaics.[260] See [30,31,261] for further information about different technologies.

2.2.3 Dynamic Hysteresis in j–V Analysis and Performance Characteristics of Perovskite Solar Cells

The analysis of j–V characteristics to assess the performance of perovskite solar cells reveals rather unusual effects: The experimentally measured j–V curve can be very sensitive to different parameters, namely scan rate, measurement direction (from forward to reverse bias or vice versa), and pre-measurement conditions, in particular, prebiasing.[217,262–266] Figure 2.9 presents a more severe example with distinct variations between the two measurement directions and the scan rate.[267] This so-called *dynamic hysteresis* effect complicates the determination of photovoltaic parameters, i.e., *open-circuit voltage* V_{oc} (at $j = 0$), *short-circuit current density* j_{sc} (at $V = 0$), *fill factor*, and power-conversion efficiency (PCE) since results vary depending on measurement conditions. Different measurement standards have been proposed and used.[262,268–271]

The so-called *maximum power-point tracking* (MPPT), nowadays one of the standards for performance characterization of perovskite solar cells, measures the maximum power-output of the device over time until its stabilization in order to determine a meaningful value for the power-conversion efficiency.[262, 269]

The observed hysteresis effects are very likely closely linked to the interplay of charge carriers and slowly moving ions (with time scales up to the order of seconds[129, 272–274]) as mentioned above. Fast j–V scans at very high scan rates could clearly separate fast carrier dynamics and slow ionic processes acting as a "slow dynamic background".[274] Numerous further studies exploring these phenomena experimentally and theoretically proposed different mechanisms[217, 273] including ferroelectric effects,[275] trapping and detrapping of charge carriers at interfaces,[276, 277] ionic motion and accumulation.[216, 278] Detailed discussions are beyond the scope of this work and can be found, e.g., in [217, 264].

Fortunately, it turned out that hysteresis effects are strongly reduced or completely vanish in most high-performance solar cells (in particular, in inverted architecture[229, 230]) and become more severe in less performing or aged devices.[193, 217, 274, 279] Since high-performance devices rely on high-quality layers and, in particular, interfaces at the charge-selective contacts this indicates the importance of their (superior) properties for (reduced) hysteresis effects.[276, 280–282] Additionally, a meaningful comparison of performance results is possible in this case (in combination with MPP tracking).

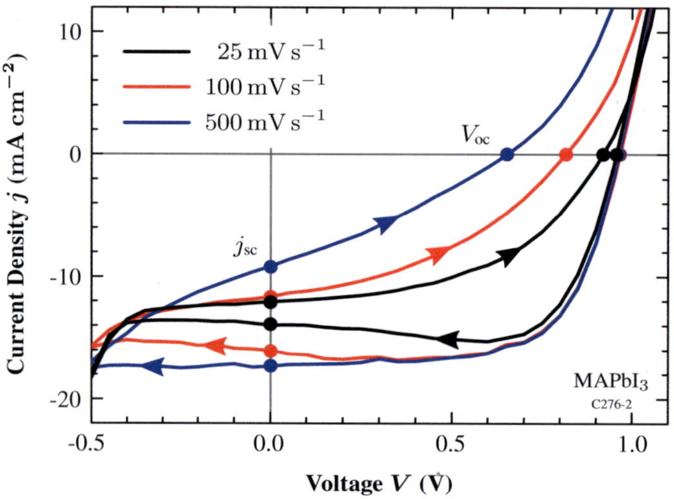

Figure 2.9: Dynamic hysteresis effect. The shape of the measured j–V curve can be strongly dependent on scan rate and measurement direction ($1.1\,\mathrm{V} \rightarrow -0.5\,\mathrm{V}$ or $-0.5\,\mathrm{V} \rightarrow 1.1\,\mathrm{V}$). Accordingly, the determination of photovoltaic parameters, i.e, V_{oc}, j_{sc}, fill factor, and PCE, is not unambigiously possible. Data obtained in collaboration with Nico Weber.[267]

Performance of Perovskite Solar Cells

Perovskite-based thin-film solar cells show remarkable performance results. The current world record of the power-conversion efficiency of 24.2% for a research cell outperforms established technologies such as $Cu(In,Ga)Se_2$ (23.3%),[14] multicrystalline silicon (22.3%),[10] and CdTe (22.1%).[10]

Values above 90% of the maximum predicted by the *Shockley–Queisser limit* (SQL)[283] have been reported for short-circuit current density j_{sc} and open-circuit voltage V_{oc} (as well as fill factors of 84% [284] and 87% [285]):

$j_{sc,max}$	$22.8\,mA\,cm^{-2}$	$\widehat{=}$	92% of SQL	with $E_g = 1.63\,eV$	[168]
	$24.9\,mA\,cm^{-2}$	$\widehat{=}$	92% of SQL	with $E_g = 1.56\,eV$	[286]
$V_{oc,max}$	$1.26\,V$	$\widehat{=}$	95% of SQL	with $E_g = 1.6\,eV$	[287]
	$1.31\,V$	$\widehat{=}$	91% of SQL	with $E_g = 1.72\,eV$	[185]

Table 2.1: Best reported values of short-circuit current density j_{sc} and open-circuit voltage V_{oc} exceed 90% of the theoretical *Shockley–Queisser limit* (SQL).

However, since these values are not achieved within one single device the power-conversion efficiencies of the best devices reach a maximum of between 70% and 80% of the Shockley–Queisser limit (compared to excellent 81% for crystalline silicon and 87% for GaAs).[185, 283, 288]

2.3 Opportunities and Challenges for Organic–Inorganic Perovskites

The exceptional optoelectronic properties of organic–inorganic perovskite materials, which led to the remarkable performance results of perovskite solar cells, also facilitate the development of promising applications even beyond photovoltaics. However, still open challenges such as toxicity and stability of the perovskite compounds and respective devices have to be overcome for successful commercialization. Both light and shade of perovskite-based applications will be summarized within this section in order to complete this review about organic–inorganic perovskites.

Tandem Solar Cells

As already mentioned above, perovskite solar cells are ideal candidates for wide-bandgap top cells in tandem solar cells. Bandgap energies of $E_g = 1.65\,eV - 1.80\,eV$ can be realized using composition tuning and represent the ideal range for a combination with silicon or

$Cu(In,Ga)Se_2$ solar cells as bottom cell ($E_g \approx 1.1$ eV).[155,156,289] Remarkable results have been achieved for both individually contacted top and bottom cells (4-terminal configuration, 4T) and monolithically stacked architectures (2-terminal devices, 2T). In case of complete thin-film tandem solar cells consisting of $Cu(In,Ga)Se_2$ and perovskite, power-conversion efficiencies up to 25.3% (4T)[151,153,154] and 22.4% (2T)[152] have been reported.

The combination with highly efficient silicon solar cells yields even higher values up to 26.4% (4T)[148] and 25.2% (2T).[57,149,150] The latter has been realized by successful deposition of the perovskite film and contact layers on top of a textured silicon cell.[150] Just recently, *OxfordPV*, a pioneering spin-out of the *University of Oxford*, set a new world record of 28% for perovskite/silicon tandem solar cells demonstrating the huge potential of this technology.[10,290,291]

Further Applications in Photovoltaics

The possible solution-based and low-temperature ($< 150\,°C$) deposition techniques (see section 2.2.2) enable fabrication of perovskite devices on flexible substrates such as polymer foils.[292–294] Additionally, especially inkjet printing[41,256,257] facilitates highly customizable perovskite solar cells which can be produced either colored[259] or with color-neutral transmission[295] for building-integrated photovoltaics.[260] Furthermore, first results on perovskite-coated flexible fibers for possible application in wearable electronics have been presented.[296,297]

Perovskite-based solar cells exhibit an unprecedented power-to-weight ratio of 23 W g^{-1} which makes them suitable candidates for photovoltaic applications in space.[298] This environment implies different requirements including low temperatures and radiation tolerance against electrons and protons.[299,300]

Perovskite Lasers and Light-Emitting Diodes

Due to the fact that high-performance solar-cell materials are also suitable for efficient light emission ("A great solar cell needs to be a great LED",[301]), light-emitting diodes (LEDs) based on organic–inorganic perovskites can been realized. The first LED operating at liquid-N_2 temperature was demonstrated years before the first solar cell.[33] Nowadays, also room-temperature performance is achieved.[34,35] Recently, the external quantum efficiency (EQE), i.e. the efficiency of electron-to-photon conversion, exceeded 20%[36,37] and, thereby, approaches the values for quantum-dot LEDs ($\sim$ 20%)[302] and OLEDs (up to 31%).[303,304] Furthermore, optically pumped amplified spontaneous emission (ASE) and lasing have been established in organic–inorganic perovskites by utilizing, e.g., distributed feedback (DFB) grating structures[38–42] or whispering-gallery mode (WGM) resonators.[43–45] In the context of light emission, perovskite nanocrystals provide promising properties in terms of photoluminescence quantum yield (PLQY), stability, and wavelength tuning and profit from the defect tolerance of perovskites.[106,160,305] Further information can be found, e.g., in [305–308].

Toxicity

Since lead is an integral constituent of most organic–inorganic perovskites questions about toxicity are raised. While some studies emphasize the toxicity of lead (and possibly methylammonium)[309, 310] others claim that the released amount of lead would be small even for complete failure of the solar-cell encapsulation and despite the water solubility of Pb^{2+}.[311] Various approaches to fabricate lead-free alternatives have been pursued but with clearly reduced power-conversion efficiency.[312–315] Tin-based perovskites yield values up to 9% but with the drawback of decreased stability and still some toxicity issues.[316–318] Other more stable compounds are $Cs_3Bi_2I_9$[319] and so-called double perovskites, e.g., $Cs_2AgBiBr_6$,[320, 321] with power-conversion efficiencies of up to 3%. Furthermore, with respect to lead-based perovskites, possible recycling of lead as well as FTO-coated substrates has been demonstrated.[322, 323]

Stability

Clearly, the major challenge for commercialization of perovskite solar cells and other perovskite-based devices is their long-term stability.[50–53] Strong progress has been made increasing the stability of perovskite solar cells from days to more than a year.[54] Some of the standard tests of the *International Electrotechnical Commission* (IEC) have been already passed,[55] e.g., thermal stability for 1000 h at 85 °C,[56] damp heat test at 85% relative humidity and 85 °C,[57, 58] and thermal cycling between −40 °C and 85 °C[58] However, additional degradation mechanisms and influences have been found which are not included in standard testing, e.g., loss of PCE during temperature cycling under operation (despite stable performance at *constant* high temperatures).[324] Overall, there is still a long way to go in understanding and preventing the various degradation mechanisms not only of the perovskite absorber but also of the charge-extraction layers (in particular, organic materials) and contacts.

One of the most detrimental influences that degrades organic–inorganic perovskites is moisture. Both reversible formation of a hydrated perovskite and irreversible degradation by formation of PbI_2 can occur depending on humidity and condensation of water.[325–327] In addition to the encapsulation of devices, the development of 2D perovskite layers on top of a 3D perovskite is a successful strategy to reduce moisture sensitivity.[54, 183–185] The presence of oxygen can also harm the perovskite or organic transport layers, in particular, in combination with UV radiation leading to the formation of O_2^- superoxides. These superoxides can easily react with the MA cation leading to formation of PbI_2 and, thereby, irreversible degradation of the perovskite.[328–330] The photocatalytic TiO_2 supports these processes and needs to be replaced by SnO_2 or optimized for long-term stability.[52, 331] Due to the possible low-temperature fabrication of perovskites and organic extraction layers, e.g., spiro-OMeTAD, these materials are also prone to thermal degradation.[332–334] Therefore, the development of stable, possibly inorganic HTLs is essential for thermal stability. Concerning the perovskite absorber itself, the exact composition can have a strong impact on ion migration that can lead to segregation in the material. For instance, halide segregation occurs under illumination after seconds or minutes in $MAPb(I_{1-x}Br_x)_3$.[335] These effects are reported to be suppressed in multiple-cation compounds including additional FA and Cs (see chapter 9 for more details).[336, 337]

2.4 Summary and Relevance for This Work

Organic–inorganic perovskites represent a class of materials crystallizing in an ABX_3 structure with an organic cation at the A site within an inorganic framework, e.g., methylammonium lead iodide, $CH_3NH_3PbI_3$, (abbreviated as $MAPbI_3$) as the most prominent example. Their specific band structure with valence band maximum and conduction band minimum both formed by antibonding molecular orbitals (in case of $MAPbI_3$, mainly from Pb–6s / I–5p and Pb–6p atomic orbitals) gives rise to high absorption coefficients and so-called defect tolerance. The resulting outstanding optoelectronic properties combined with possible low-temperature and solution-processed fabrication entail the successful application of perovskite materials in thin-film solar cells, LEDs, lasers and many more.

Despite remarkable power-conversion efficiencies of perovskite solar cells exceeding 24%,[10] fundamental understanding lacks behind in some areas. One example is the binding energy E_B of excitons (i.e., bound electron–hole pairs) in perovskites which is still under debate.[79, 338–341] The determination of E_B for different perovskite compounds is one of the objectives of this work (see chapter 7). Furthermore, the influence of excitonic effects on fundamental transitions in $MAPbI_3$ are investigated using temperature-dependent optical spectroscopy. In the context of solar cells, these questions are of particular relevance since the separation of photogenerated electron–hole pairs is the fundamental working principle in photovoltaics. Therefore, implications of excitonic effects on solar-cell performance are discussed.

The feasibility of compositional engineering for optimization of desired material properties, mainly bandgap tuning and stability enhancement, is an important aspect for further evolution of perovskite devices. Multiple-cation lead mixed-halide compounds including Cs, FA, and MA as well as I and Br have demonstrated promising properties in terms of improved power-conversion efficiency and stability. Therefore, a special emphasis of this work is on the optical and structural properties of these compounds. On the one hand, temperature-induced phase transitions between different crystal phases (orthorhombic and tetragonal) and their impact on the bandgap energy are studied (see chapter 8). The occurence of phase transitions in the temperature range of operation of solar cells or lasers can have a significant impact on the performance. Furthermore, stability of organic–inorganic perovskites and related devices is the key challenge for commercialization. Mixed-halide compounds show fast segregation of the halides under illumination which can be reduced by additional cation incorporation. The stability of the promising multiple-cation lead mixed-halide compounds is thoroughly examined in chapter 9. Electromodulation spectroscopy provides the required sensitivity for determination of small changes in bandgap energy which serve as an indicator for segregation effects in this work.

Therefore, the utilized absorption and electromodulation spectroscopy are presented in the following chapters. Since the dielectric function describes the interaction of light and matter it represents the underlying foundation for the understanding of spectroscopic techniques and is thus introduced in chapter 3.

Chapter 3

Implications of Excitons on Dielectric Function and Absorption in Semiconductors

Bound electron–hole pairs, named *excitons*, can play an important role in semiconducting materials such as organic–inorganic perovskites, especially for the application in photovoltaics. By absorption of photons in the semiconductor, simultaneously electrons in the conduction band and holes in the valence band are generated. The efficient extraction of these charge carriers is crucial for the performance of the photovoltaic device. Consequently, the formation of bound electron–hole pairs can hinder separation and extraction of charge carriers in solar cells. Therefore, the analysis of excitonic effects in absorber layers of solar cells is of particular relevance.

This chapter serves as an introduction on how excitons affect the optical properties of direct semiconductors. Since all optical characteristics are embedded in the dielectric function ϵ, an overview of the basic modeling of ϵ will be given as well as important fundamental relationships, e.g. the so-called *Kramers–Kronig relations* (section 3.1). After a discussion of optical excitations in the one-electron picture neglecting charge-carrier interactions, the attractive Coulomb potential between negatively charged electrons and positively charged holes is implemented leading to two-particle bound states named excitons (section 3.2). Equipped with the modeling of the wave function and energy levels of weakly bound *Wannier–Mott excitons*, their impact on the dielectric function as well as the respective *absorption coefficient* is discussed. The resulting *Elliott formula* describes the absorption spectrum including excitonic effects and demonstrates significant changes not only due to bound states, but also because of a continuum of ionized excitonic states above the bandgap energy (section 3.3).

3.1 The Dielectric Function of Semiconductors

The essential foundation of every optical spectroscopy is the interaction of matter with electromagnetic waves. In the framework of Maxwell's equations, this is described by the fundamental relation $\boldsymbol{D} = \epsilon_0 \boldsymbol{E} + \boldsymbol{P}$ of electric displacement field $\boldsymbol{D}$, electric field $\boldsymbol{E}$ and polarization field $\boldsymbol{P}$.

For the materials and excitation densities used in this work, the discussion can be restricted to the case of linear optics, for which a linear response of the polarization $\boldsymbol{P}$ induced in the material by an external electric field $\boldsymbol{E}$ is assumed. This is characterized by the susceptibility χ or the commonly used dielectric function ϵ, which, in general, depend on frequency ω and wave vector $\boldsymbol{k}$.[342–344]

$$\frac{1}{\epsilon_0}\boldsymbol{P}(\omega,\boldsymbol{k}) = \chi(\omega,\boldsymbol{k})\,\boldsymbol{E}(\omega,\boldsymbol{k}) = \left(\epsilon(\omega,\boldsymbol{k}) - 1\right)\boldsymbol{E}(\omega,\boldsymbol{k}) \tag{3.1}$$

Generally being a second-rank tensor, $\epsilon(\omega,\boldsymbol{k})$ is reduced to a scalar function for isotropic media. Since the wavelength of light is much larger than relevant atomic distances, e.g., lattice constant or exciton radius, in most cases, the $\boldsymbol{k}$ dependence of $\epsilon(\omega,\boldsymbol{k})$ (so-called *spatial dispersion*) can be neglected.[343, 344] The remaining $\epsilon(\omega)$ is a complex function $\epsilon(\omega) = \epsilon_\mathrm{r}(\omega) + \mathrm{i}\epsilon_\mathrm{i}(\omega)$.

The real and imaginary part of the dielectric function are linked by so-called *Kramers–Kronig relations* (in the absence of magnetic fields)[342–344]

$$\begin{aligned}\epsilon_\mathrm{r}(\omega) - 1 &= \frac{2}{\pi}\,\mathrm{P}\int_0^\infty \frac{\omega'\epsilon_\mathrm{i}(\omega')}{\omega'^2-\omega^2}\mathrm{d}\omega' \\ \epsilon_\mathrm{i}(\omega) &= \frac{2\omega}{\pi}\,\mathrm{P}\int_0^\infty \frac{\epsilon_\mathrm{r}(\omega')}{\omega^2-\omega'^2}\mathrm{d}\omega',\end{aligned} \tag{3.2}$$

where P denotes the Cauchy principal part of the integral. These fundamental relationships are solely based on causality since the response to an electric field cannot appear before the electric field itself. Moreover, they imply that either $\epsilon_\mathrm{r}(\omega)$ or $\epsilon_\mathrm{i}(\omega)$ are sufficient for the description of optical material properties.

The exact modeling of the dielectric function $\epsilon(\omega)$ requires a detailed understanding of possible excitations in the system. In case of transitions from the valence to the conduction band in direct semiconductors, first-order perturbation theory (*Fermi's Golden Rule*) yields, e.g., for the imaginary part of the dielectric function[345]

$$\epsilon_\mathrm{i}(\omega) \propto \frac{1}{\omega^2}\int_\mathrm{BZ} \mathrm{d}^3\boldsymbol{k}\,\left|\boldsymbol{P}_\mathrm{cv}(\boldsymbol{k})\right|^2 \delta\left(E_\mathrm{c}(\boldsymbol{k}) - E_\mathrm{v}(\boldsymbol{k}) - \hbar\omega\right) \tag{3.3}$$

where $|\boldsymbol{P}_\mathrm{cv}(\boldsymbol{k})|$ designates the matrix element for transitions from the valence band $E_\mathrm{v}(\boldsymbol{k})$ to the conduction band $E_\mathrm{c}(\boldsymbol{k})$ using the dipole approximation and $\int_\mathrm{BZ}$ the integral over the first Brillouin zone. The Dirac delta function δ ensures energy conservation and can be alternatively written in the Fourier-transformed time representation resulting in[78, 343, 346]

$$\epsilon_\mathrm{i}(\omega) \propto \frac{1}{\omega^2}\int_\mathrm{BZ} \mathrm{d}^3\boldsymbol{k}\,\left|\boldsymbol{P}_\mathrm{cv}(\boldsymbol{k})\right|^2 \int_{-\infty}^{\infty} \exp\left[-\frac{\mathrm{i}}{\hbar}\left(E_\mathrm{c}(\boldsymbol{k}) - E_\mathrm{v}(\boldsymbol{k}) - \hbar\omega\right)t\right]\mathrm{d}t \tag{3.4}$$

which will be further used for the discussion of electric-field induced effects on ϵ in chapter 5.2. Since the transition matrix element is approximately constant close to the band edge, this result can simplified by introducing the $\boldsymbol{k}$-independent *oscillator strength* $f_\mathrm{cv} \propto |\boldsymbol{P}_\mathrm{cv}|/\omega$ and performing the integration over $\boldsymbol{k}$ which leads to[342, 343]

$$\epsilon_\mathrm{i}(\omega) \propto \frac{1}{\omega}f_\mathrm{cv}\cdot D_\mathrm{j}(\hbar\omega)\,. \tag{3.5}$$

The remaining integration over the constant energy surface $S_{\boldsymbol{k}}$ (with $E_{\mathrm{cv}} = \mathrm{const.}$) defines the *joint density of states* (JDOS)[342, 343, 345]

$$D_{\mathrm{j}}\left(E_{\mathrm{cv}}\right) = \frac{1}{4\pi^3} \int_{S_{\boldsymbol{k}}} \frac{\mathrm{d}S_{\boldsymbol{k}}}{\left|\nabla_{\boldsymbol{k}}\left(E_{\mathrm{cv}}\right)\right|}, \tag{3.6}$$

which essentially describes the density of optical transitions for a given energy difference $E_{\mathrm{cv}} := E_{\mathrm{cv}}(\boldsymbol{k}) = E_{\mathrm{c}}(\boldsymbol{k}) - E_{\mathrm{v}}(\boldsymbol{k})$ of conduction and valence band (Accordingly, $\hbar\omega = E_{\mathrm{cv}}$ for optical transitions described in $\epsilon(\omega)$.).

In other words, the imaginary part ϵ_{i} of the dielectric function is determined by the combination of the oscillator strength f_{cv}, describing the transition probability, as well as the density of possible transitions D_{j}. Accordingly, in the vicinity of the band edge, the dielectric function $\epsilon_{\mathrm{i}}(\omega)$ is mainly governed by the shape of the JDOS.

The singularities in $D_{\mathrm{j}}(E_{\mathrm{cv}})$ for vanishing $\left|\nabla_{\boldsymbol{k}}(E_{\mathrm{cv}})\right|$ (see (3.6)) are named *critical points* or *Van-Hove singularities* and are classified in M_0, M_1, M_2, and M_3 (see [342–344] for details). The energy of critical points in the band structure can be precisely determined by modulation spectroscopy which will be discussed in chapter 5.

For an M_0 critical point, e.g., the fundamental absorption edge of a direct semiconductor, the calculation of $D_{\mathrm{j}}(E_{\mathrm{cv}} = \hbar\omega)$ in the vicinity of the band edge results in a square-root dependence $\propto \sqrt{\hbar\omega - E_{\mathrm{g}}}$ above the bandgap energy E_{g} (assuming parabolic band dispersion and 3 dimensions). Consequently, the imaginary part ϵ_{i} of the dielectric function is described by[343, 345]

$$\epsilon_{\mathrm{i}}(\omega) = \begin{cases} \frac{\mathrm{const.}}{\omega^2}\sqrt{\hbar\omega - E_{\mathrm{g}}} & \hbar\omega \geq E_{\mathrm{g}} \\ 0 & \hbar\omega \leq E_{\mathrm{g}} \end{cases} \tag{3.7}$$

The corresponding absorption coefficient α can be deduced from the imaginary part of the dielectric function ϵ_{i} via[343–345]

$$\alpha(\omega) = 2k_0 \cdot \kappa(\omega) = \frac{\omega}{c_0 \cdot n(\omega)} \cdot \epsilon_{\mathrm{i}}(\omega), \tag{3.8}$$

where $\tilde{n}(\omega) = n(\omega) + \mathrm{i}\kappa(\omega)$ denotes the complex refractive index and is linked to $\epsilon(\omega)$ by $\epsilon_{\mathrm{r}} = n^2 - \kappa^2$ and $\epsilon_{\mathrm{i}} = 2n\kappa$. c_0 and k_0 are the speed and the wavenumber of light in vacuum, respectively. By using (3.5) and (3.7), the absorption coefficient can be rewritten as[343, 345]

$$\alpha(\hbar\omega) = \frac{\mathrm{const.}}{nc_0} f_{\mathrm{cv}} \cdot D_{\mathrm{j}}(\hbar\omega) = \begin{cases} \frac{\mathrm{const.}}{\hbar\omega n}\sqrt{\hbar\omega - E_{\mathrm{g}}} & \hbar\omega \geq E_{\mathrm{g}} \\ 0 & \hbar\omega \leq E_{\mathrm{g}} \end{cases} \tag{3.9}$$

Equivalent to $\epsilon_{\mathrm{i}}(\omega)$, the absorption coefficient of a direct semiconductor close to the band edge is strongly dominated by the square-root dependence of the JDOS. Taking finite lifetime broadening Γ into account smears out the absorption onset at the bandgap energy E_{g} and leads to the expression $\alpha(\hbar\omega) \propto \sqrt{\hbar\omega - E_{\mathrm{g}} + \mathrm{i}\Gamma}$ (similar is true for $\epsilon_{\mathrm{i}}(\omega)$).[346, 347]

However, so far we only used the one-electron approximation for the description of direct band-to-band transitions. In fact, multi-particle excitations such as excitons play a crucial role in many semiconductors, e.g., organic-inorganic perovskites.

3.2 Fundamentals of Excitons

In the one-electron picture, we only considered the excitation of a single electron from the valence to the conduction band leading to a "missing" electron in the valence band, a so-called hole. However, since electrons and holes are charged (quasi-)particles, they influence each other via the Coulomb interaction. This attractive potential leads to a correlation of the motion of electron and hole and, as a result, to bound electron–hole pairs, in analogy to the hydrogen or positronium atom, which are called *excitons*.[342–344, 348]

Since the Coulomb potential only depends on the relative position of electron and hole to each other, it is convenient to decompose the exciton motion into a relative motion of electron and hole, and a center-of-mass translation. In most semiconductors, the Coulomb potential is strongly screened by valence electrons (associated with large dielectric constants) leading to only weakly interacting electrons and holes.[343–345] These weakly bound electron–hole pairs are known as *Wannier–Mott excitons* (or simply *Wannier excitons*) and can be described in the effective mass approximation with m_e and m_h for electron and hole, respectively. The resulting problem for the relative motion of electron and hole can be solved in analogy to the hydrogen atom. The dispersion relation of excitons (consisting of relative and center-of-mass contribution) is then given by[343, 344]

$$E_\mathrm{ex}\left(n_\mathrm{B}, \boldsymbol{K}\right) = E_\mathrm{g} - E_\mathrm{B}\frac{1}{n_\mathrm{B}^2} + \frac{\hbar^2 \boldsymbol{K}^2}{2M} \tag{3.10}$$

with the principal quantum number $n_\mathrm{B} = 1, 2, 3, ...$, center-of-mass wave vector $\boldsymbol{K} = \boldsymbol{k}_\mathrm{e} + \boldsymbol{k}_\mathrm{h}$ and mass $M = m_\mathrm{e} + m_\mathrm{h}$ of the exciton. The *exciton binding energy* E_B of the bound electron–hole pair (also known as *excitonic Rydberg energy* following the hydrogen atom) and the related *exciton Bohr radius* a_B^ex are defined as[344, 348]

$$E_\mathrm{B} = \mathrm{Ry}\frac{\mu}{m_0}\frac{1}{\epsilon^2} \tag{3.11}$$

$$a_\mathrm{B}^\mathrm{ex} = a_\mathrm{B}\epsilon\frac{m_0}{\mu} \tag{3.12}$$

with reduced exciton mass $\mu = \frac{m_\mathrm{e} m_\mathrm{h}}{m_\mathrm{e}+m_\mathrm{h}}$, Rydberg energy $\mathrm{Ry} = 13.6\,\mathrm{eV}$ and Bohr radius $a_\mathrm{B} \approx 0.053\,\mathrm{nm}$ of the hydrogen atom and electron mass m_0.[348]

In typical semiconductors, the exciton binding energy E_B is in the range of some meV, e.g., 4.9 meV in GaAs[343] or 14.7 meV in Si,[349] up to 59 meV in ZnO,[343, 344] due to reduced μ and large ϵ because of the screened Coulomb potential. Similarly, the exciton Bohr radius is significantly increased compared to the hydrogen atom. Typical values are in the range of $1\,\mathrm{nm} - 100\,\mathrm{nm}$ and, thus, much larger than the lattice constant, which justifies the effective mass approximation for *Wannier excitons* in a self-consistent way.[343, 344, 348] However, excitons which are confined to one unit cell exist in some materials, e.g., alkaline halides such as NaCl, and are called *Frenkel excitons*.[343–345]

The dispersion relation of Wannier excitons is depicted in Fig. 3.1. As described by (3.10), it consists of a series of parabolas describing the center-of-mass dispersion. For $n_\mathrm{B} \to \infty$, the excitonic states merge into the ionization continuum above the bandgap energy E_g.

The wave function of Wannier excitons $\Psi^{\boldsymbol{K}}_{n_\mathrm{B},l,m}(\boldsymbol{r}_\mathrm{e}, \boldsymbol{r}_\mathrm{h})$ consists of the conduction- and valence-band functions $\phi_\mathrm{e,h}(\boldsymbol{r}_\mathrm{e,h})$ with space coordinate $\boldsymbol{r}_\mathrm{e,h}$ of electron or hole, respectively, the plane-wave factor $e^{\mathrm{i}\boldsymbol{K}\cdot\boldsymbol{R}}$ describing the free propagation of the exciton through the crystal and an envelope function $\phi^\mathrm{env}_{n_\mathrm{B},l,m}(\boldsymbol{r})$:[344, 350]

$$\Psi^{\boldsymbol{K}}_{n_\mathrm{B},l,m}(\boldsymbol{r}_\mathrm{e}, \boldsymbol{r}_\mathrm{h}) = \sqrt{\Omega}\ \phi_\mathrm{e}(\boldsymbol{r}_\mathrm{e})\,\phi_\mathrm{h}(\boldsymbol{r}_\mathrm{h})\,e^{\mathrm{i}\boldsymbol{K}\cdot\boldsymbol{R}}\phi^\mathrm{env}_{n_\mathrm{B},l,m}(\boldsymbol{r}) \tag{3.13}$$

with normalization factor $\sqrt{\Omega}$ and the relative and center-of-mass coordinates $\boldsymbol{r} = \boldsymbol{r}_\mathrm{e} - \boldsymbol{r}_\mathrm{h}$ and $\boldsymbol{R} = \frac{m_\mathrm{e}\boldsymbol{r}_\mathrm{e} + m_\mathrm{h}\boldsymbol{r}_\mathrm{h}}{m_\mathrm{e} + m_\mathrm{h}}$, respectively. The envelope functions $\phi^\mathrm{env}_{n_\mathrm{B},l,m}(\boldsymbol{r})$ are solutions to the modified hydrogen problem and describe the relative motion of electron and hole.[344, 350] By using the wave function $\Psi^{\boldsymbol{K}}_{n_\mathrm{B},l,m}(\boldsymbol{r}_\mathrm{e}, \boldsymbol{r}_\mathrm{h})$ in (3.13), the influence of Wannier excitons on the absorption can be derived to study excitonic semiconductors such as organic–inorganic perovskites.

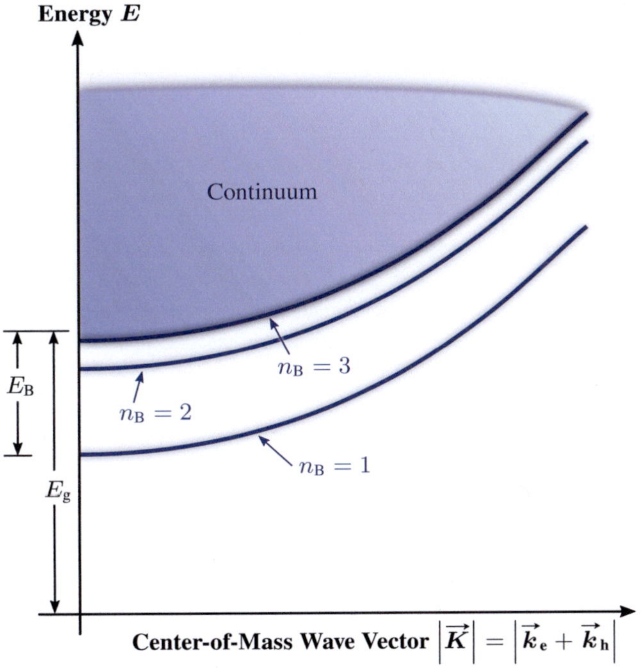

Figure 3.1: Dispersion relation of Wannier excitons in a semiconductor with bandgap energy E_g and exciton binding energy E_B (see (3.10)). Bound states with $n_\mathrm{B} = 1, 2, 3$ and continuum states are illustrated. Visualization based on [344].

3.3 Absorption of Excitonic Semiconductors and Elliott Formula

In order to evaluate the influence of excitons on the absorption in the vicinity of the band edge of a direct semiconductor (in general: in the vicinity of a M_0 critical point), the conversion probability of photons into excitons can be calculated by using Fermi's Golden Rule, similarly to (3.3):[343]

$$W_{0f} \propto \sum_{f} |\boldsymbol{P}_{0f}|^2 \ \delta\left(E_f(\boldsymbol{K}) - E_0 - \hbar\omega\right), \tag{3.14}$$

where $|\boldsymbol{P}_{0f}|^2$ describes the transition matrix element from ground state with no excitons and energy E_0 to the final excited state with an exciton with energy E_f and wave vector $\boldsymbol{K}$. Due to momentum conservation, $\boldsymbol{K}$ must be equal to the negligible photon wave vector leading to $K \approx 0$. Calculations by Elliott demonstrate that the transition matrix element $|\boldsymbol{P}_{0f}|^2$ and, thereby, the oscillator strength f_{n_B} of the n_B-th excitonic transition is directly proportional to the square of the envelope function $\phi^{env}_{n_B,l,m}$ at $\boldsymbol{r} = 0$:[77, 343]

$$f_{n_B} \propto |\boldsymbol{P}_{0f}|^2 \propto \left|\phi^{env}_{n_B,l,m}(\boldsymbol{r}=0)\right|^2. \tag{3.15}$$

Since $\boldsymbol{r} = 0$ is identical to $\boldsymbol{r}_e = \boldsymbol{r}_h$, this can be interpreted in such a way that f_{n_B} is determined by the probability of finding a conduction-band electron and a valence-band hole within the same lattice unit cell or, in other words, by the overlap of the electron and hole wave function.[78, 343]

The further derivation of the imaginary part of the dielectric function $\epsilon_i(\omega)$ and the corresponding absorption $\alpha(\hbar\omega)$ leads to an expression consisting of two contributions: a series of discrete absorption lines due to bound electron–hole pairs and a continuum of ionized excitons above the bandgap. Both are presented in the following sections.

3.3.1 Bound States of Electrons and Holes

According to the dispersion relation in (3.10), for vanishing wave vector $\boldsymbol{K}$, excitons exhibit a series of discrete energy levels $\propto 1/n_B^2$ below the bandgap energy E_g which show up in the absorption spectrum as discrete lines.

Because $\left|\phi^{env}_{n_B,l,m}(0)\right|^2$ and, thereby, the oscillator strength f_{n_B} are nonzero exclusively for $l = 0$ (analogous to the hydrogen atom), only excitons with s-like wave functions can be optically excited. By inserting the respective envelope functions, this leads to[77, 344]

$$f_{n_B} \propto \frac{1}{n_B^3}. \tag{3.16}$$

Similar to (3.5), the imaginary part of the dielectric function can be expressed by[78,79,343]

$$\epsilon_{\mathrm{i}}(\omega) \propto \sum_{n_{\mathrm{B}}=1}^{\infty} \frac{1}{n_{\mathrm{B}}^{3}} \, \delta\left(\hbar\omega - E_{\mathrm{g}} + \frac{E_{\mathrm{B}}}{n_{\mathrm{B}}^{2}}\right), \tag{3.17}$$

where the δ function serves a similar purpose as the JDOS in (3.5). Accordingly, $\epsilon_{\mathrm{i}}(\omega)$ and the corresponding absorption spectrum consist of a sum of infinitely sharp peaks with a decreasing intensity $\propto 1/n_{\mathrm{B}}^{3}$.

3.3.2 Ionization Continuum Above the Bandgap Energy

In the ionization continuum for energies $\hbar\omega > E_{\mathrm{g}}$, the excitons are ionized and electrons and holes are no longer in a bound state. However, there is still a residual effect of the Coulomb interaction due to the mutual attraction of electrons and holes. This leads to a correlation of their statistical positions in space and, thereby, an enhanced wavefunction overlap. Analogously to (3.15), the oscillator strength is determined by the probability of finding electron and hole in the same unit cell. Calculations by Elliott yield[77,78,342,343]

$$\left|\phi^{\mathrm{env}}_{n_{\mathrm{B}},l,m}(\boldsymbol{r}=0)\right|^{2} = \frac{\tau e^{\tau}}{NV \sinh(\tau)}, \tag{3.18}$$

where N is the number of unit cells, V the volume of one unit cell and $\tau = \pi\sqrt{\frac{E_{\mathrm{B}}}{\hbar\omega - E_{\mathrm{g}}}}$.

The density of states for excitons with parabolic dispersion is similar to (3.7) given by[77]

$$D(\hbar\omega) \propto \sqrt{\hbar\omega - E_{\mathrm{g}}} \tag{3.19}$$

Following an equivalent argument as for (3.5), the imaginary part of the dielectric function results in[343]

$$\epsilon_{\mathrm{i}}(\omega) \propto \frac{\sqrt{\hbar\omega - E_{\mathrm{g}}}}{\omega^{2}} \cdot \frac{\tau e^{\tau}}{\sinh(\tau)}. \tag{3.20}$$

For vanishing exciton binding energy $E_{\mathrm{B}} \rightarrow 0$ leading to $\tau \rightarrow 0$, $\frac{\tau e^{\tau}}{\sinh(\tau)}$ is close to unity and reproduces the non-excitonic case in (3.7).

3.3.3 Elliott Formula and Sommerfeld Enhancement Factor

Combining the contributions of bound electron–hole pairs and the ionization continuum in (3.17) and (3.20), and calculating the corresponding absorption $\alpha(\hbar\omega)$ results in the so-called *Elliott formula*[77, 79, 351]

$$\alpha(\hbar\omega) \propto \frac{\mu_{\mathrm{cv}}^2}{\hbar\omega}\left[\underbrace{\sum_{n_{\mathrm{B}}=1}^{\infty}\frac{4\pi\sqrt{E_{\mathrm{B}}^3}}{n_{\mathrm{B}}^3}\,\delta\left(\hbar\omega - E_{\mathrm{g}} + \frac{E_{\mathrm{B}}}{n_{\mathrm{B}}^2}\right)}_{\text{Contribution of bound excitons}} + \underbrace{\theta(\hbar\omega - E_{\mathrm{g}})\,\frac{\pi\sqrt{E_{\mathrm{B}}}\,e^{\tau}}{\sinh(\tau)}}_{\text{Ionization continuum}}\right], \tag{3.21}$$

with transition matrix element μ_{cv}^2 (similar to $|\boldsymbol{P}_{\mathrm{cv}}|^2$ above, e.g., in (3.3)) and the Heaviside step function θ.

This result is also confirmed by more elaborate calculations by Haug and Koch.[78] Figure 3.2 illustrates the absorption spectrum predicted by the Elliott formula.

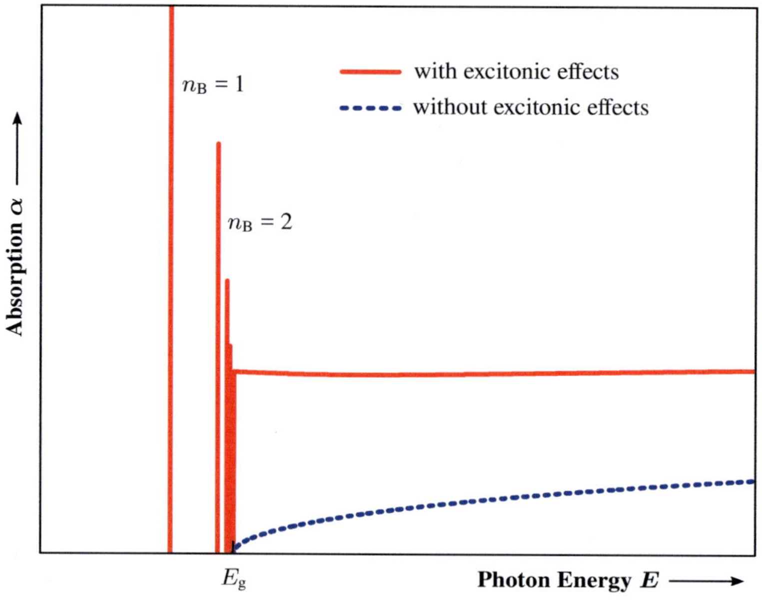

Figure 3.2: Elliott Formula (red curve) consisting of a series of discrete lines arising from bound electron–hole pairs and the ionization continuum above the bandgap energy E_{g} (Intensities of discrete lines are not to scale but should decrease with $1/n_{\mathrm{B}}^3$.). The dashed blue line shows the non-excitonic, square-root-like absorption.

The comparison of the excitonic continuum absorption $\alpha_{\text{continuum}}$ (see (3.20) and (3.8)) and the absorption without excitonic effects $\alpha_{\text{no excitons}}$ in (3.9) yields the so-called *Sommerfeld enhancement factor*[78, 348]

$$S = \frac{\alpha_{\text{continuum}}}{\alpha_{\text{no excitons}}} = \frac{\tau e^{\tau}}{\sinh(\tau)} = \frac{2\pi\sqrt{\frac{E_{\text{B}}}{\hbar\omega - E_{\text{g}}}}}{1 - \exp\left(-2\pi\sqrt{\frac{E_{\text{B}}}{\hbar\omega - E_{\text{g}}}}\right)}. \tag{3.22}$$

For photon energies $\hbar\omega$ close to the bandgap energy E_{g}, S is approximately constant to $1/\sqrt{\hbar\omega - E_{\text{g}}}$ compensating the square-root dependence in the non-excitonic case. This results in a constant absorption close to the band edge and demonstrates the significant influence of excitonic effects due to the Coulomb interaction even above the bandgap energy E_{g} (see Fig. 3.2). For vanishing exciton binding energy E_{B}, S is close to unity and the first contribution of bound excitons in (3.21) vanishes, recovering the result for free charge carriers in (3.9).

This discussion of the *Sommerfeld enhancement* is continued in chapter 7.3.1 with a more illustrative description of excitonic effects by means of the example of perovskites.

3.4 Summary

This chapter presents an overview about the optical properties of semiconductors starting from the dielectric function ϵ and highlighting the influence of bound electron–hole pairs, i.e., excitons, on ϵ as well as on the absorption spectrum.

The complex dielectric function ϵ contains all information about the optical properties of a solid-state material. Due to causality, its real and imaginary part are fundamentally linked by Kramers–Kronig relations which will be used for the evaluation of experimental spectra in chapter 5.

For semiconductors, the imaginary part ϵ_{i} can be modeled by calculating the transition rate of electrons from the valence to the conduction band due to the absorption of photons using Fermi's Golden Rule. The resulting absorption spectrum mainly depends on the joint density of states (JDOS) of valence and conduction band and results in a square-root dependence in the vicinity the bandgap. Additionally, the JDOS exhibits singularities at critical points which can be experimentally determined using modulation spectroscopy (see chapter 5).

However, for a complete description of semiconducting materials, bound states of electrons and holes due to the attracting Coulomb potential need to be considered. These electron–hole pairs are only weakly bound in most semiconductors because of the screened Coulomb potential and are called Wannier–Mott excitons with discrete energy levels similar to the hydrogen atom.

By taking these two-particle excitations into account, the resulting absorption spectra are significantly changed. Below the bandgap energy E_{g}, a series of discrete lines arises from the discrete excitonic energy levels. However, even above E_{g}, the absorption spectrum is modified by the correlation of electron and hole movement. The change of this ionization continuum compared to the square-root dependence in the non-excitonic case is described by the Sommerfeld enhancement factor S. This results in an constant absorption closely above the bandgap. The complete description of the absorption including excitons is combined in the so-called Elliott formula.

Still, for an effective application in the evaluation of experimental absorption spectra, further extensions are necessary which will be presented in the following chapter.

Chapter 4

Absorption Spectroscopy and Evaluation Methods

The detailed description of the impact of excitonic effects on the absorption coefficient of semiconductors by the Elliott formula presented in chapter 3 facilitates a verification by experimentally obtained absorption spectra. This chapter presents the experimental realization (section 4.1) and necessary background for the evaluation of measured spectra. The latter includes phenomenological modifications of the Elliott formula such as finite *line-width broadening* Γ to enable least-squares fitting to experimental data (section 4.2). In addition, the concept of the so-called *f-sum rule method* proposed by Sestu *et al.*[352] can be utilized for determination of the exciton binding energy E_B from temperature-dependent absorption spectra (section 4.3). It relies on the fact that the oscillator strength f is independent of the temperature-dependent line-width broadening Γ and can be employed for a more reliable evaluation of E_B by normalization and integration of measured spectra.

4.1 Experimental Realization of Temperature-Dependent Absorption Spectroscopy

The absorption of a material can be calculated based on energy conservation of the incident light:[353]

$$I_{\text{incident}} = I_{\text{reflection}} + I_{\text{transmission}} + I_{\text{absorption}} \qquad \Leftrightarrow \qquad 1 = R + T + A' \tag{4.1}$$

which is also reflected by the relative coefficients $\{R, T, A'\} = I_x / I_{\text{incident}}$ of reflected, transmitted, and absorbed light intensity $I_x = \{I_{\text{reflection}}, I_{\text{transmission}}, I_{\text{absorption}}\}$.

Generally, for the evaluation of absorption spectra, not the so-called *absorptance* A' but the logarithmic *absorbance* A is used which is, for transparent samples ($R \approx 0$), given by[353]

$$A \approx -\log_{10} T. \tag{4.2}$$

It is related to the absorption coefficient α by $A = (d/\ln 10) \cdot \alpha$ with d being the distance of the light traveling through the medium.[353]

In order to obtain absolute values for absorptance A' and absorbance A, both reflectance R and transmittance T are measured using an *Ulbricht sphere* (or *integrating sphere*) to collect also scattered light. However, this is not feasible for temperature-dependent measurements using a liquid-helium cryostat. Still, for the evaluation of absorbance spectra performed in this work, the assumptions of transparent samples and negligible scattering are sufficient (similar to most reports in literature[353]). An exemplary comparison of measurement results with spectra obtained by a commercial spectrophotometer with integrating sphere (*PerkinElmer Lambda 1050*) at room temperature is presented in Fig. A.1 (a) in appendix A.1.

Figure 4.1 depicts the experimental setup to record temperature-dependent absorption spectra. Since the bandgap energy E_g of the investigated perovskites is in the range of $\sim 1.5\,\mathrm{eV} - 2.3\,\mathrm{eV}$ ($\sim 830\,\mathrm{nm} - 540\,\mathrm{nm}$) a halogen lamp serves as suitable illumination source. The light is guided onto the sample by a series of lenses (L) and a pinhole (PH). The detection of the transmitted light is performed using a grating monochrometer with 0.275 m focal length following a charge-coupled device (CCD) camera. For the wavelength calibration a neon spectral lamp is used. The sample is mounted in a liquid-helium cryostat which allows to control the temperature T in a range of 4.2 K up to room temperature. The transmission coefficient T and the related absorbance A are calculated utilizing a reference spectrum recorded without the sample.

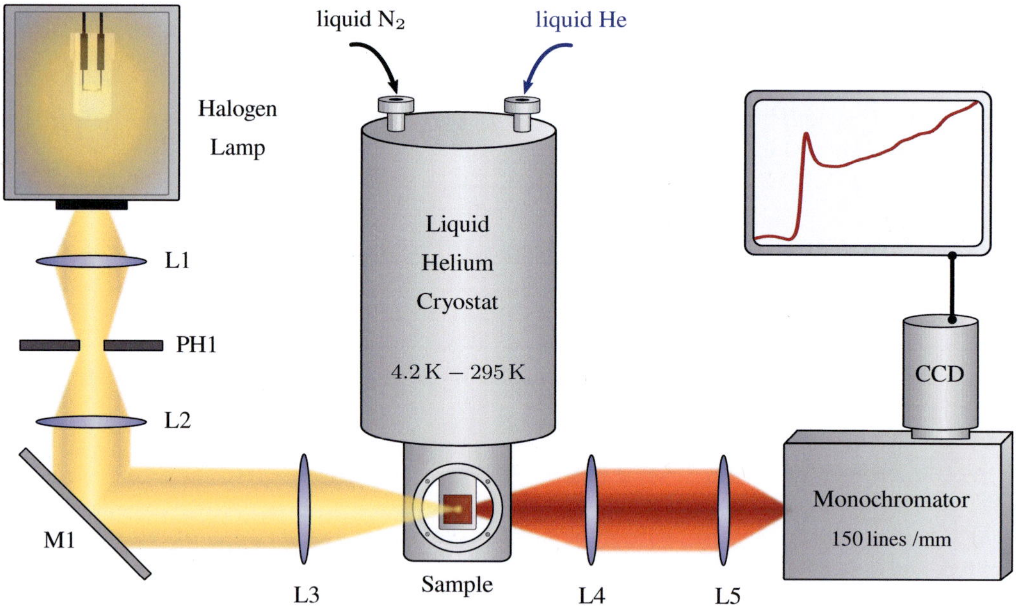

Figure 4.1: Experimental realization of temperature-dependent absorption spectroscopy within this work. The sample is mounted in a liquid-helium cryostat ($T = 4.2\,\mathrm{K} - 295\,\mathrm{K}$) and illuminated by a halogen lamp. The transmitted light is focused on the entrance slit of a grating monochromator and detected by a CCD camera. Detailed information about setup components can be found in appendix A.3.

Further details about each component of the setup can be found in appendix A.3. Additionally, a different setup was used for the simultaneous measurement of absorption and modulation spectra which is presented in chapter 6.1.

4.2 Generalized Elliott Formula for the Evaluation of Absorption Spectra

For an effective application of the Elliott formula in the evaluation of experimental spectra, further phenomenological advancements are necessary. Obviously, the infinitesimal line width of the bound excitonic states needs to be replaced by a realistic finite line width Γ which accounts, e.g., for homogeneous lifetime broadening. This can be realized by substituting the δ function in (3.21) with a bell-shaped function such as the hyperbolic secant function $\mathrm{sech}\,(x/\,\Gamma)$.[78,79,351] Similar can be achieved for the ionization continuum by convolution with $\mathrm{sech}\,(x/\,\Gamma)$.

Additionally, Saba *et al.* introduced a correction factor b in the joint dispersion relation of valence and conduction band to account for deviations from the assumption of perfectly parabolic bands[79]

$$E_\mathrm{c}\left(\boldsymbol{k}\right) - E_\mathrm{v}\left(\boldsymbol{k}\right) = E_\mathrm{g} + \frac{\hbar^2 \boldsymbol{k}^2}{2\mu} - b\boldsymbol{k}^4 \tag{4.3}$$

with the reduced electron–hole mass μ. The resulting joint density of states $D'\left(\hbar\omega\right)$ is then given by

$$D'\left(\hbar\omega\right) \propto \frac{1}{1 - \frac{8\mu^2 b}{\hbar^4}\left(\hbar\omega - E_\mathrm{g}\right)}\sqrt{\hbar\omega - E_\mathrm{g}} \tag{4.4}$$

which reproduces the parabolic assumption for $b = 0$.

A *generalized version* of the *Elliott formula* including finite *line-width broadening* Γ and *non-parabolicity correction factor* b is then suitable for the evaluation of measured absorption spectra:[79]

$$\alpha\left(\hbar\omega\right) \propto \frac{\mu_\mathrm{cv}^2}{\hbar\omega}\left[\underbrace{\sum_{n_\mathrm{B}=1}^{\infty} \frac{2\sqrt{E_\mathrm{B}^3}}{n_\mathrm{B}^3}\,\mathrm{sech}\left(\frac{\hbar\omega - E_{\mathrm{n_B}}^\mathrm{B}}{\Gamma}\right)}_{\text{Sum of discrete exciton peaks}} + \underbrace{\int_{E_\mathrm{g}}^{\infty} \mathrm{sech}\left(\frac{\hbar\omega - E}{\Gamma}\right)\frac{\sqrt{E_\mathrm{B}}}{1 - \exp\left(-2\pi\sqrt{\frac{E_\mathrm{B}}{E - E_\mathrm{g}}}\right)} \cdot \frac{1}{1 - \frac{8\mu^2 b}{\hbar^4}\left(E - E_\mathrm{g}\right)}\,\mathrm{d}E}_{\text{Continuum contribution including Sommerfeld enhancement}}\right] \tag{4.5}$$

Figure 4.2 illustrates the influence of different values for Γ and b. Larger values of the non-parabolicity factor b lead to an increased upwards-bending of the initially constant ionization continuum at higher photon energies. Finite line-width broadening Γ results in a broadening of both the discrete excitonic peaks and the onset of the ionization continuum. Correspondingly, the sum of excitonic peaks is strongly dominated by the $n = 1$ contribution and higher-order peaks can only be resolved for $\Gamma \lesssim 0.1\,E_\mathrm{B}$.

The *generalized Elliott formula* enables the evaluation of experimental spectra using least-squares fits. However, a careful analysis of these *Elliott fits* is required because of numerous fitting parameters.

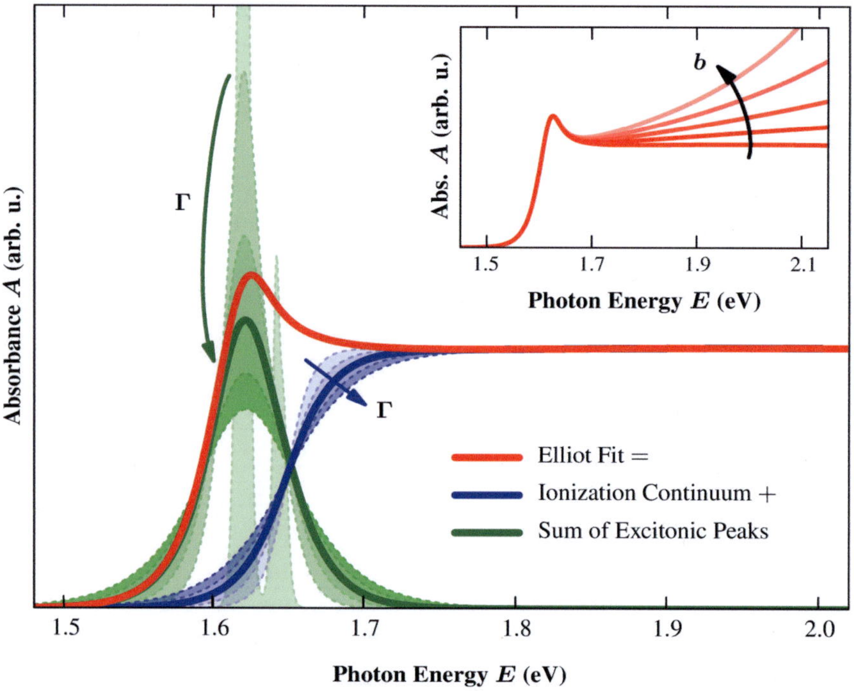

Figure 4.2: Illustration of the generalized Elliott formula (red) proposed by Saba *et al.* consisting of a sum of discrete excitonic peaks (green) and the ionization continuum (blue).[79] Realistic parameters for $MAPbI_3$ are used ($E_\mathrm{g} = 1.65\,\mathrm{eV}$, $E_\mathrm{B} = 30\,\mathrm{meV}$, $\Gamma = 20\,\mathrm{meV}$) but with ideally parabolic bands ($b = 0$). Finite line-width broadening Γ acts on both the width of the excitonic peaks and the onset of the continuum (here $\Gamma = 2\,\mathrm{meV} - 30\,\mathrm{meV}$). The sum of the broadened excitonic peaks is strongly dominated by the $n = 1$ states. Only for very small $\Gamma \lesssim 0.1\,E_\mathrm{B}$, the $n = 2$ peak can be resolved which is typically not observed in absorption measurements of perovskites. Inset: A finite non-parabolicity factor b leads to an upward-bending of the constant ionization continuum (here $\frac{8\mu^2 b}{\hbar^4} = 0\,\mathrm{eV}^{-1} - 1.2\,\mathrm{eV}^{-1}$).

4.3 f-Sum Rule Method for Determination of the Exciton Binding Energy

The analysis of temperature-dependent absorption spectra by fitting the generalized Elliott formula in (4.5) to experimental data yields reliable results for the bandgap energy E_g and the exciton binding energy E_B, especially at cryogenic temperatures. However, for temperatures T approaching room temperature, the reliability can be compromised, in particular for semiconductors with small E_B, because the increasing line-width broadening Γ can lead to difficulties in resolving the sum of discrete excitonic peaks and the ionization continuum separately. While the accuracy of the determined bandgap energy E_g might still be sufficient for the desired application, the situation is much more critical for exciton binding energies E_B in the meV range. A representative example of these difficulties is shown in Fig. B.1 in appendix B.

In order to improve the reliability of the determination of E_B, Sestu *et al.* proposed a different method.[352] The basic idea is illustrated in Fig. 4.3 based on calculated spectra using the generalized Elliott formula in (4.5). Afterwards the underlying principle will be translated into a practical evaluation method.

All curves depicted in Fig. 4.3 (a) and (b) are shifted by the respective energy position $E_\mathrm{ex} = E_\mathrm{g} - E_\mathrm{B}$ of the excitonic peak maximum which facilitates better comparison. Additionally, each spectrum is normalized to the absorbance at a defined energy E_n (in this case $E_\mathrm{n} = E_\mathrm{ex} + 0.25\,\mathrm{eV}$). The normalization eliminates any dependence on the prefactor μ_cv^2 in (4.5) which is typically influenced by lattice contraction or expansion as well as phase transitions.[352]

In Fig. 4.3 (a), the broadening parameter Γ is varied while E_B is kept constant. Thereby, the shape of the excitonic continuum is not strongly modified, but only the excitonic peak and the band edge become broadened (see also Fig. 4.2). The integral I over the depicted energy range, the so-called *integrated normalized absorption*, which is defined as[352]

$$I = \frac{1}{\alpha\left(E_\mathrm{n}\right)} \int \alpha\left(E\right) dE \tag{4.6}$$

stays approximately constant in the case of varying Γ at fixed E_B which is presented in Fig. 4.3 (c) as dashed blue curve.

In contrast, if Γ is kept constant while varying E_B (Fig. 4.3 (b)) both the discrete lines and the ionization continuum are clearly affected. The corresponding integrated normalized absorption I is no longer constant but increases with E_B as shown in Fig. 4.3 (c) as solid red curve.

The underlying physical principle behind this phenomenon is a so-called *f-sum rule*. The total activity of oscillators (or resonances) in this frequency range (in general, all frequencies) is given by the sum of all oscillators described by their oscillator strength f.[78,345] This *f-sum* is independent of any details of the applied model or the dielectric function as long as all oscillators are included.[78,345,352]

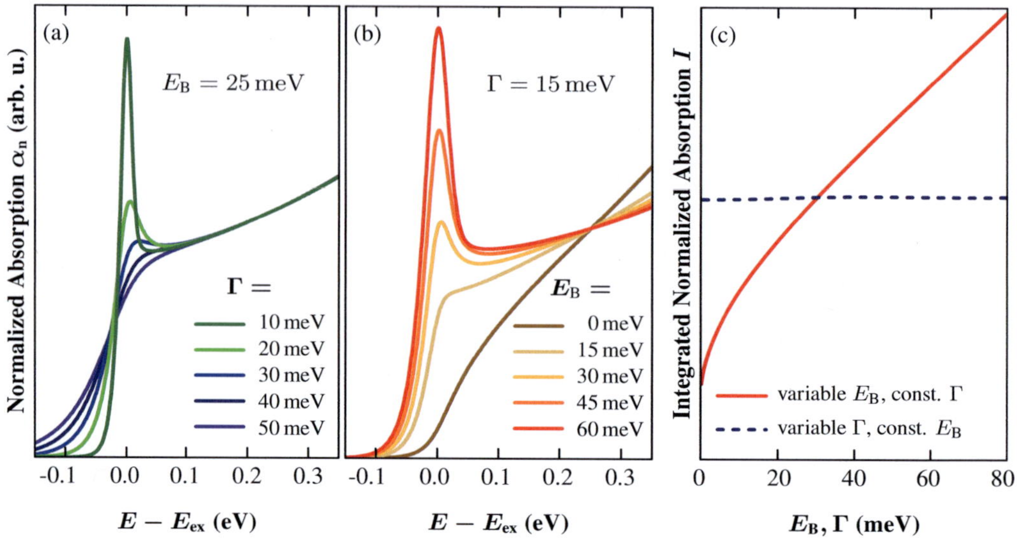

Figure 4.3: Illustration of the basic principle behind the f-sum rule method proposed by Sestu *et al.*[352] The spectra are calculated according to the generalized Elliott formula in (4.5), shifted by the excitonic peak position E_{ex} and normalized to $E_n = E_{ex} + 0.25\,\mathrm{eV}$ for better comparison. Parameters similar to $MAPbI_3$ are used ($E_g = 1.6\,\mathrm{eV}$, $E_B = 25\,\mathrm{meV}$ in (a), and $\Gamma = 15\,\mathrm{meV}$ in (b)). (a) The variation of Γ only affects the excitonic features close to the absorption edge. (b) Varying E_B strongly affects both the discrete lines and the excitonic continuum. (c) The integrated normalized absorption I is nearly constant for varying Γ (dashed blue line) as in (a) but increases for larger values of E_B (red curve) as in (b). This is based on the fact, that the oscillator strength f does only change with E_B but not with Γ, leading to a so-called f-sum rule. Visualization based on [352].

Since the line-width broadening Γ only defines the (spectral) spread of the transitions (or oscillators) contributing to the absorption, the integral I of all transitions is independent of Γ and stays constant.[352] On the other hand, a change of the exciton binding energy E_B modifies the oscillator strength and, thereby, the integral I.

This principle is exploited in the proposed f-sum rule method since for known values of the non-parabolicity factor b, the integral becomes a function $I(E_B)$ which only depends on E_B and serves as a direct measure for the exciton binding energy.

In practice, experimental absorption spectra are shifted by E_{ex} (determined by a fit of the excitonic peak or linear interpolation for completely washed-out peaks at high temperatures) and normalized to the absorbance $\alpha(E_n)$ at E_n. For the subsequent numerical integration of the experimentally obtained spectra, the normalization energy E_n is also chosen as upper limit of the integrated normalized absorption I_{exp} (with index "exp" for the experimental spectra). Thus, the choice of E_n influences the value of I_{exp} but not the f-sum rule as long as $E_n - E_g \gg \Gamma$. However, the best sensitivity of I_{exp} on E_B is expected when E_n leads to comparable contributions of the discrete and continuum part in the integral.[352]

Additionally, the lower bound of the integral in (4.6) is chosen to be $E_{ex} - \Delta$ with Δ being large enough to ensure that the whole excitonic peak is included even for large Γ at room temperature. Simultaneously, Δ needs to be small enough to reject any low-energy absorption which is not due to band-to-band transitions.[352]

In order to obtain quantitative values for the exciton binding energy E_B, the function $I(E_B)$ for the integrated normalized absorption needs to be inverted. Since this is not analytically possible, the "inversion" is achieved by calculating $I(E_B)$ as in Fig. 4.3 (c) utilizing the generalized Elliott formula and inserting the parameters used in the experimental evaluation, e.g, normalization energy E_n and lower bound of the integral.[352] By comparison with the experimental value I_{exp}, the integrated normalized absorption can be mapped onto E_B.

The remaining parameter $(8\mu^2 b)/\hbar^4$ describing non-parabolic bands needs to be obtained by directly fitting the generalized Elliott formula to absorption data. Since the determination of $(8\mu^2 b)/\hbar^4$ does not rely on the (eventually washed-out) excitonic peak but on the upwards-bending at higher photon energy, it allows for a robust evaluation even at higher temperatures. Sestu *et al.* reported negligible variations of the non-parabolicity factor with temperature for $MAPbI_3$ and $MAPbBr_3$.[352] The rather weak temperature dependence of $(8\mu^2 b)/\hbar^4$ found in this work is shown in Fig. B.2. Furthermore, implications for the reliability of the f-sum rule method at high temperatures compared to direct Elliott fits are discussed in appendix B.

The f-sum rule method proposed by Sestu *et al.* can help to increase the reliability of the obtained values for the exciton binding energy at high temperatures approaching room temperature where the excitonic peak in the absorption spectrum can no longer be clearly resolved by the Elliott formula.

4.4 Summary

The experimental methods for measuring temperature-dependent absorption spectra presented in this chapter are straightforward and well-established in semiconductor spectroscopy.

Nevertheless, a careful evaluation and analysis is required in order to obtain reliable results. For a direct least-squares fit of the Elliott formula presented in chapter 3 phenomenological modifications are necessary: Line-width broadening is taken into account by the convolution with a bell-shaped function with width Γ. Deviations from the assumption of parabolic bands close to the bandgap are considered by an additional $b\boldsymbol{k}^4$ term in the joint dispersion relation of valence and conduction band with non-parabolicity factor b. The resulting generalized Elliott formula facilitates the evaluation of measured absorption spectra by least-squares fitting in order to determine the bandgap energy E_g and the exciton binding energy E_B.

However, especially for large line-width broadening Γ in the order of E_B at temperatures approaching room temperature, the reliability and accuracy of the fit can be reduced, mainly influencing the determination of E_B. The so-called f-sum rule method proposed by Sestu *et al.*[352] relies on the fact that the sum of the oscillator strength f is independent of the line-width broadening Γ. Therefore, the integral I of the normalized absorption can be used as a universal measure for the exciton binding energy E_{B}. Still, this method implies that the temperature dependence of the non-parabolicity $\mu^2 b/\hbar^4$ is weak within the investigated temperature range and does not affect I. Thus, with careful analysis of temperature-dependent absorption spectra, the f-sum rule method can improve the reliability of determined exciton binding energies E_{B} even up to room temperature.

Chapter 5

Modulation Spectroscopy of Semiconductors: Free Charge Carriers and Excitonic Effects

Combined with detailed fitting and analysis techniques, absorption spectroscopy provides a well applicable and powerful tool to investigate electronic properties, e.g., bandgap energy or exciton binding energy, which are embedded in the dielectric function. However, depending on the properties of interest of the dielectric function, even more suitable methods are available. In particular, this is true for precise investigations of critical points in the band structure (introduced in chapter 3.1, see especially (3.6)) for which modulation spectroscopy is the method of choice.

This chapter presents fundamentals of modulation spectroscopy including objective and basic concept as well as theoretical foundation and physical background. After a brief introduction in modulation spectroscopy techniques in general (section 5.1), the focus remains on electromodulation spectroscopy for the rest of this work since it is the most suitable type for the application to solar cells. Therefore, in the following chapter, the influence of an electric field on the dielectric function of a direct semiconductor is discussed in the framework of *Franz–Keldysh theory* as well as Aspnes' approximations in the *low-field regime* (section 5.2).[346, 347] A special emphasis is on the scenario of confined and, in particular, excitonic states (section 5.2.3). The respective line shapes for the measurable optical spectra, i.e., reflectance and transmittance, are presented for both non-excitonic band-to-band transitions and excitonic resonances (section 5.3). Additionally, experimental spectra can be evaluated by a so-called *modulus transformation* based on Kramers–Kronig relations.[73, 354] Finally, this chapter concludes with a comparison of absorption and electromodulation spectroscopy highlighting the respective advantages and possible applications.

5.1 Basic Principle of Modulation Spectroscopy

Characteristic features in the dielectric function and the corresponding optical quantities, reflectance and transmittance, are spectrally rather broad, especially at room temperature. Van-Hove singularities, e.g., at the bandgap energy E_g (see (3.6) and chapter 3.1) are of particular interest for the examination of the electronic structure in semiconductors, but appear only weakly in these spectra compared to the broad background. However, the derivatives of the optical quantities with respect to energy diverge at these points.[343, 346, 355, 356] Modulation spectroscopy allows to directly access these derivatives experimentally. The obtained sharp derivative-like resonance spectra enable a precise detection of critical points while strongly suppressing any spectrally broad background.[343, 346, 355, 356]

The fundamental concept of modulation spectroscopy techniques is always based on the detection of small changes in the optical spectrum of a sample while modulating either the measurement system (*internal modulation*) or the physical properties of the material (*external modulation*).[346, 356] The relative changes of reflectance $\Delta R/R$ or transmittance $\Delta T/T$ are typically in the order of $10^{-3} - 10^{-6}$, which can be measured with phase-sensitive detection using a lock-in amplifier.[343, 346, 356] The normalization with respect to the absolute value R or T entails the benefit of eliminating any dependence on detector response or intensity fluctuations of the light source.[342, 343, 346, 356] Modulation of the measurement system can be performed by modifying the excitation wavelength. On the other hand, external modulation is achieved by application of a perturbation in temperature (*thermomodulation*), stress (*piezomodulation*), illumination intensity (*photomodulation*), magnetic or electric field (*electromodulation*).[342, 343, 346, 356]

The different types of external modulation always affect the band structure and, thereby, modify the joint density of states which enters the dielectric function (see, e.g., (3.5) and (3.6)). Therefore, a variation of the electronic structure translates into a variation of the dielectric function and, in turn, a change of reflectance and transmittance. Depending on the type of perturbation, the physical mechanism of modulation can be divided into two categories: modulations which preserve lattice periodicity, e.g., temperature or stress, and perturbations which break the translation symmetry of the crystal, e.g., electric and magnetic fields.[342, 347] In the former case, optical transitions remain vertical due to momentum conservation and negligible photon momentum as illustrated in Fig. 5.1 (a). The modulation mainly affects the energy, broadening or amplitude of the respective optical transition and can often be described by first-derivative line shapes.[347] On the other hand, electric and magnetic fields break the translation symmetry of the crystal lattice and, thereby, momentum conservation along the field direction. Consequently, the vertical optical transitions are washed out and the resulting line shapes can only be described by higher-order derivatives (see Fig. 5.1 (b)).[347]

In order to obtain theoretically expected line shapes, the experimental $\Delta R/R$ or $\Delta T/T$ spectra need to be linked to the change of the dielectric function. Since the modulation amplitudes are always small, a linear relation is justified:[59, 343, 346, 356, 357]

$$\frac{\Delta R}{R}\left(\text{or } \frac{\Delta T}{T}\right) = \alpha\left(\epsilon_\mathrm{r}, \epsilon_\mathrm{i}\right)\Delta\epsilon_\mathrm{r} + \beta\left(\epsilon_\mathrm{r}, \epsilon_\mathrm{i}\right)\Delta\epsilon_\mathrm{i}. \tag{5.1}$$

The so-called *Seraphin coefficients* (here exemplary for R)

$$\alpha\left(\epsilon_\mathrm{r}, \epsilon_\mathrm{i}\right) = \frac{\partial \ln R}{\partial \epsilon_\mathrm{r}} \quad \text{and} \quad \beta\left(\epsilon_\mathrm{r}, \epsilon_\mathrm{i}\right) = \frac{\partial \ln R}{\partial \epsilon_\mathrm{i}} \tag{5.2}$$

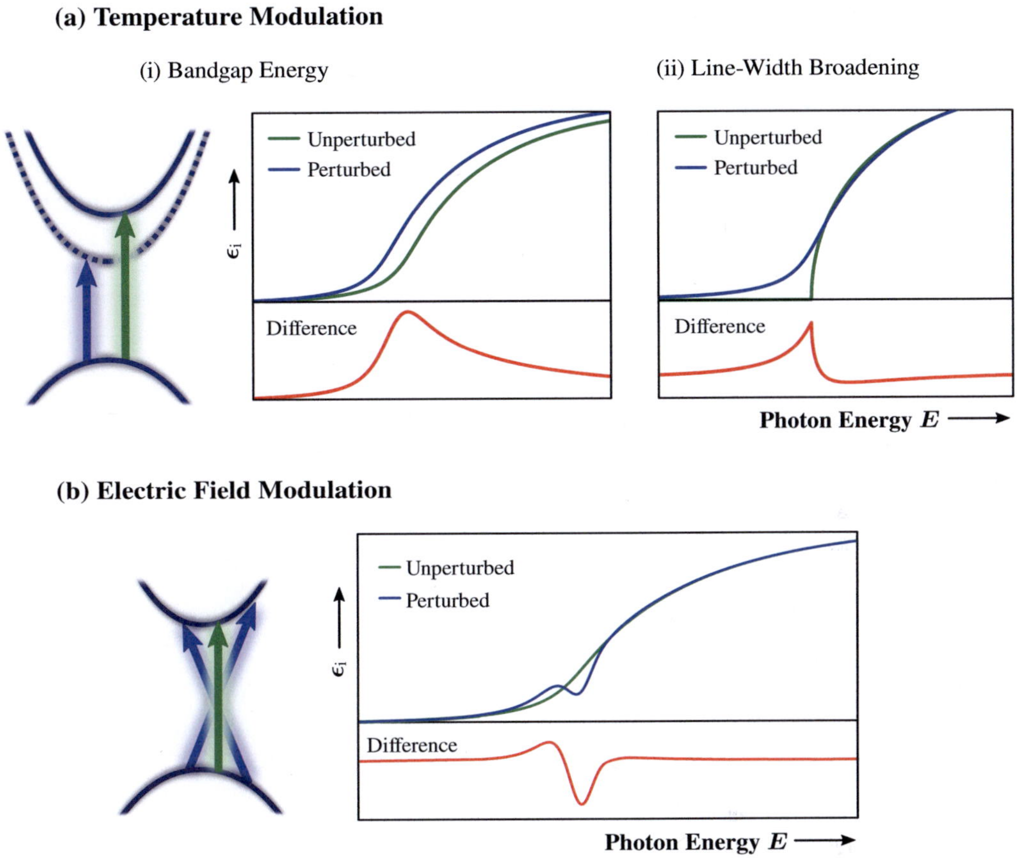

Figure 5.1: The underlying physical mechanisms of modulation spectroscopy for a perturbation which (a) conserves or (b) breaks the translation symmetry of the periodic lattice. Temperature modulation as an example for (a) can cause modifications of (i) the bandgap energy or (ii) the lifetime broadening both resulting in a first-derivative line shape. In contrast to (a), the modulation by an electric field allows non-vertical transitions since momentum is no longer conserved and free charge-carriers can be accelerated. See also [342, 347].

are obtained by differentiation of the Fresnel expressions for reflectance R or transmittance T.[59,346] The change in the dielectric function ϵ_r and ϵ_i requires a detailed modeling of the impact of the applied perturbation. In the scope of this work, electromodulation spectroscopy is used because of its facile applicability especially to solar cells, even for multilayer thin-film devices,[74–76,241,341,358] and well-defined resonance features in experimental spectra. Consequently, the influence of electric fields on the dielectric function of semiconductors with and without pronounced excitonic effects and the implications for the measured line shapes are discussed in the following sections.

5.2 Influence of Electric Fields on the Dielectric Function of Semiconductors

While the basic experimental realization of electromodulation spectroscopy is rather straightforward, the correct evaluation and interpretation of the measured spectra can be challenging and require a deeper understanding of the modulation mechanism. For this reason, the influence of electric fields on the dielectric function of a direct-bandgap semiconductor will be reviewed in this section. First fundamental work on this topic has been performed by Franz and Keldysh after which the corresponding effect is named.[359,360] Aspnes, Pollak and many more studied the implications on electromodulation in detail and developed practical models, e.g., for the low-field regime.[346,347,355,356,361–366]

5.2.1 Franz–Keldysh Theory

The *Franz–Keldysh theory* describes the influence of an electric field $\boldsymbol{\mathcal{E}}$ on unbound electrons and holes in a semiconductor and the resulting alterations of the optical properties. The electric field destroys the translational symmetry of the solid-state material along the field direction and accelerates unbound electrons and holes. The corresponding implications on the dielectric function can be treated either by a stationary scalar potential or by using time-dependent pertubation theory. The former can be found in [361, 362].

Here, the calculation of the transition rate is presented similar to (3.3) which results in an expression for the imaginary part of the dielectric function ϵ_i of the form[346,347,363]

$$\epsilon_\mathrm{i}(\omega) \propto \frac{1}{\omega^2} \int_\mathrm{BZ} \mathrm{d}^3\boldsymbol{k}\, |\boldsymbol{P}_\mathrm{cv}(\boldsymbol{k})|^2 \int_{-\infty}^{\infty} \mathrm{d}s \exp\left[\mathrm{i}\left(\hbar\omega + \mathrm{i}\Gamma - \int_{-s/2}^{s/2} \mathrm{d}s' E_\mathrm{cv}(\boldsymbol{k} - e\boldsymbol{\mathcal{E}}s')\right)s\right], \tag{5.3}$$

where Γ denotes the line-width broadening and $E_\mathrm{cv}(\boldsymbol{k}) = E_\mathrm{c}(\boldsymbol{k}) - E_\mathrm{v}(\boldsymbol{k})$ the joint energy dispersion of valence and conduction band. $s^{(\prime)} = t^{(\prime)}/\hbar$ (with time t) is introduced for clarity.

The accelerating impact of the electric field on unbound electrons is reflected by a momentum change of[346]

$$\hbar\frac{\mathrm{d}\boldsymbol{k}}{\mathrm{d}t} = -e\boldsymbol{\mathcal{E}}. \tag{5.4}$$

caused by the electrostatic force. The related additional wave vector $-e\boldsymbol{\mathcal{E}}t/\hbar$ "smears" out the formerly vertical transitions (due to negligible photon momentum) in the band structure leading to also non-vertical optical transitions (see Fig. 5.1 (b)). For $\boldsymbol{\mathcal{E}} = 0$, (5.3) recovers (3.3) (here with vanishing $\Gamma = 0$).

In case of sufficiently weak electric fields, the joint energy dispersion can be expanded in a Taylor series[363]

$$E_{\mathrm{cv}}\left(\boldsymbol{k}-e\boldsymbol{\mathcal{E}}s'\right)=E_{\mathrm{cv}}\left(\boldsymbol{k}\right)-es'\left(\boldsymbol{\mathcal{E}}\cdot\boldsymbol{\nabla}_{\boldsymbol{k}}\right)E_{\mathrm{cv}}\left(\boldsymbol{k}\right)+\frac{e^2s'^2}{2}\left(\boldsymbol{\mathcal{E}}\cdot\boldsymbol{\nabla}_{\boldsymbol{k}}\right)^2E_{\mathrm{cv}}\left(\boldsymbol{k}\right)-\ldots \tag{5.5}$$

which is truncated after the quadratic term. Since the symmetric integral of the linear contribution vanishes, this results in[346, 347, 363]

$$\epsilon_{\mathrm{i}}\left(\omega\right)\propto\frac{1}{\omega^2}\int_{\mathrm{BZ}}\mathrm{d}^3\boldsymbol{k}\,\left|\boldsymbol{P}_{\mathrm{cv}}\left(\boldsymbol{k}\right)\right|^2\int_{-\infty}^{\infty}\mathrm{d}s\,\exp\left[\mathrm{i}\left(\hbar\omega+\mathrm{i}\Gamma-E_{\mathrm{cv}}\left(\boldsymbol{k}\right)\right)s-\mathrm{i}\frac{1}{3}\left(\hbar\Omega\right)^3s^3\right] \tag{5.6}$$

The so-called *electro-optic energy* (defined as $\hbar\Omega$ or $\hbar\theta$ in literature)[346, 347, 363]

$$\left(\hbar\Omega\right)^3=\frac{1}{8}e^2\left(\boldsymbol{\mathcal{E}}\cdot\boldsymbol{\nabla}_{\boldsymbol{k}}\right)^2E_{\mathrm{cv}}\left(\boldsymbol{k}\right)=\frac{1}{4}\frac{e^2\mathcal{E}^2\hbar^2}{2\mu_{\|}}=\frac{1}{4}\left(\hbar\theta\right)^3. \tag{5.7}$$

is the characteristic energy of a particle with mass $\mu_{\|}$ in a uniform electric field $\boldsymbol{\mathcal{E}}$. $\mu_{\|}$ denotes the interband reduced mass of valence and conduction band along the electric-field direction which is constant under the assumption of parabolic bands.

Neglecting the broadening parameter Γ, (5.6) can be simplified by using the definition of the *Airy function* of the first kind[346, 363]

$$\mathrm{Ai}\left(x\right)=\frac{1}{2\pi}\int_{-\infty}^{\infty}\mathrm{d}s\,\exp\left(\frac{1}{3}\mathrm{i}s^3+\mathrm{i}xs\right) \tag{5.8}$$

which yields the following expression for the imaginary part of the dielectric function:[346, 363]

$$\epsilon_{\mathrm{i}}\left(\omega\right)\propto\frac{1}{\omega^2}\int_{\mathrm{BZ}}\mathrm{d}^3\boldsymbol{k}\,\left|\boldsymbol{P}_{\mathrm{cv}}\left(\boldsymbol{k}\right)\right|^2\underbrace{\left[\frac{1}{\hbar\Omega}\mathrm{Ai}\left(\frac{E_{\mathrm{cv}}\left(\boldsymbol{k}\right)-\hbar\omega}{\hbar\Omega}\right)\right]}_{=\delta\left(E_{\mathrm{cv}}\left(\boldsymbol{k}\right)-\hbar\omega\right)\text{ for }\hbar\Omega\to0} \tag{5.9}$$

The physical interpretation of the Airy function is the mixing of different contributions in the vicinity of $\boldsymbol{k}$. Accordingly, in the limit of vanishing electric fields $\boldsymbol{\mathcal{E}}$, i.e., $\hbar\Omega\to0$, the Airy function is replaced by the Dirac delta distribution $\delta\left(E_{\mathrm{cv}}\left(\boldsymbol{k}\right)-\hbar\omega\right)$ and the form of (3.3) is restored.[363] This is synonymous to the recovery of energy conservation due to translational symmetry.

The remaining mathematical calculations do not provide deeper insights into physical principles and can be found in [361]. The final result of these derivations and approximations describes $\Delta\epsilon$ in the so-called *intermediate field regime* as classified by Aspnes.[347] Additionally, it should be noted that a more elaborate approach using the complex version of the Airy function enables the integration of line-width broadening Γ.

The resulting solution for the intermediate field regime is given by[367]

$$\Delta\epsilon_{\mathrm{r}} = B\left(\hbar\theta\right)^{1/2}\mathrm{Im}\left[\frac{H\left(z\right)}{\left(\omega - \mathrm{i}\Gamma\right)^{2}}\right] \tag{5.10a}$$

$$\Delta\epsilon_{\mathrm{i}} = B\left(\hbar\theta\right)^{1/2}\mathrm{Re}\left[\frac{H\left(z\right)}{\left(\omega - \mathrm{i}\Gamma\right)^{2}}\right], \tag{5.10b}$$

with

$$z = \frac{\left(E_{\mathrm{g}} - \hbar\omega\right) + \mathrm{i}\Gamma}{\hbar\theta}. \tag{5.11}$$

and the complex *electro-optic function*

$$H\left(z\right) = \pi\left[\mathrm{Ai}'^{2}\left(z\right) - z\mathrm{Ai}^{2}\left(z\right)\right] + \mathrm{i}\left(\left[\mathrm{Ai}'\left(z\right)\mathrm{Bi}'\left(z\right) - z\mathrm{Ai}\left(z\right)\mathrm{Bi}\left(z\right)\right] + z^{1/2}\right). \tag{5.12}$$

$\mathrm{Ai}\left(z\right)$ and $\mathrm{Bi}\left(z\right)$ are the complex Airy functions of first and second kind, respectively. The prefactor B includes the transition matrix element.[346, 367]

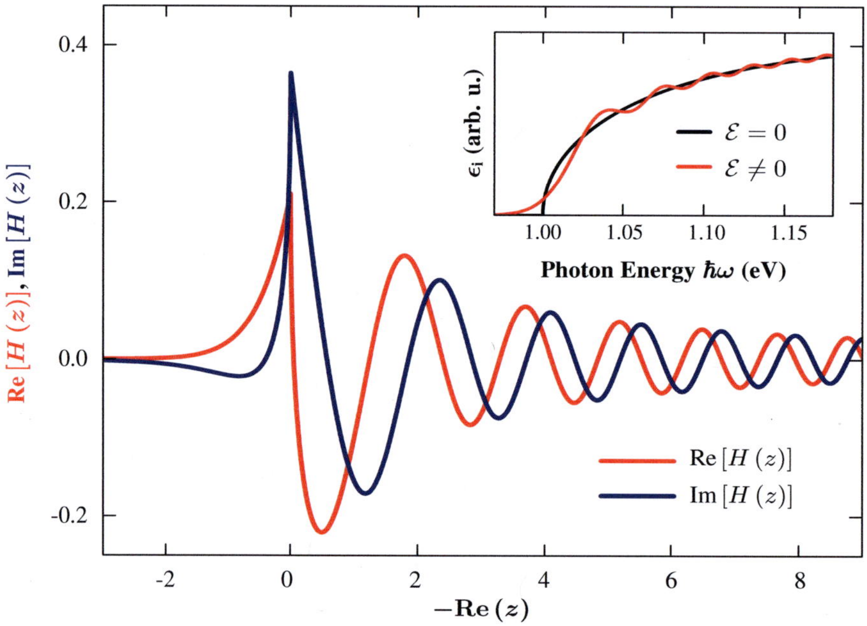

Figure 5.2: Illustrations of real (red) and imaginary (blue) part of the electro-optic function $H\left(z\right)$ for $\Gamma = 0$. The inset shows the imaginary part $\epsilon_{\mathrm{i}}\left(\omega\right)$ of the dielectric function without (black) and with (red) an electric field (with $\Gamma = 0$, $E_{\mathrm{g}} = 1\,\mathrm{eV}$). For photon energies above the bandgap energy E_{g}, i.e., $\mathrm{Re}\left(z\right) < 0$, Franz–Keldysh oscillations (FKO) appear. The exponential tail below E_{g} ($\mathrm{Re}\left(z\right) > 0$) can be explained by the tunneling of electrons from valence to conduction band due to tilting of the bands by the electric field. Visualization based on [343].

Figure 5.2 depicts the real and imaginary part of $H(z)$ (often denoted as electro-optic functions $F(z)$ and $G(z)$ in literature).[343,346,367] For $\mathrm{Re}(z) < 0$, i.e., photon energies $\hbar\omega$ above the bandgap energy E_g, characteristic so-called Franz–Keldysh oscillations appear which are also visible in the dielectric function.[346,356,363] For photon energies $\hbar\omega < E_\mathrm{g}$ ($\mathrm{Re}(z) > 0$), the dielectric function is no longer zero as in the case without applied electric field but decays exponentially.

In order to understand the physical origin of these effects, a slightly different perspective can be helpful for further insights: The application of an electric field leads to an additional linear potential $e\mathcal{E}z$ in space along the field direction z which causes a spatial tilt of the bands along z. Thereby, strictly speaking, the energy gap is removed since for a certain (distance) value of z both valence and conduction band are on the same energy level. Accordingly, the exponential tail below E_g can be seen as a result of electrons tunneling a distance z from valence to conduction band.[342,343,356]

An alternative explanatory approach considers the spatial-dependent overlap of electron and hole wave functions in conduction and valence band, respectively. Both are mathematically described by Airy functions which consist of an oscillatory part and decay exponentially in the spatial region between the (spatially) tilted conduction and valence band. Due to the linear potential $e\mathcal{E}z$ electron and hole (Airy) wave functions are shifted relative to each other depending on the energy difference. The resulting change of the wave function overlap can explain both the exponential decay below E_g and the oscillatory behavior above E_g. Further details on this approach can be found in [368,369].

The physical phenomena as well as the mathematical derivation of the intermediate field regime discussed in this section rely on the condition of an unchanged band structure, i.e., $e\mathcal{E}a \ll E_\mathrm{g}$ (with real space length a of a unit cell), justifying the use of pertubation theory. On the other hand, the electro-optic energy $\hbar\theta$ can exceed the broadening Γ which leads to clearly observable Franz–Keldysh oscillations.[347] The following section will introduce the so-called *low-field regime* for $\hbar\theta < \Gamma$.[346,347]

5.2.2 The Low-Field Regime

The Franz–Keldysh oscillations are effectively "washed" out for increasing lifetime broadening Γ compared to the electro-optic energy $\hbar\theta$ which is visualized in Fig. 5.3.

For sufficiently small electric fields $\mathcal{E}$ and, thereby, small $\hbar\theta$ and $\hbar\Omega$, the lifetime broadening term $e^{-\Gamma s}$ cuts off the integrand in (5.6) before higher-order terms play a significant role justifying the approximation[346,347,357]

$$\exp\left[-\mathrm{i}\frac{1}{3}(\hbar\Omega)^3 s^3\right] \approx 1 - \mathrm{i}\frac{1}{3}(\hbar\Omega)^3 s^3. \tag{5.13}$$

The corresponding condition for the low-field limit is given by $\hbar\Omega \leq \Gamma/3$.[370, 371] Accordingly, the change of the dielectric function can be expressed by

$$\Delta\epsilon \propto -\mathrm{i}\frac{1}{3}(\hbar\Omega)^3 \frac{1}{(\hbar\omega)^2}\int_{\mathrm{BZ}} \mathrm{d}^3\boldsymbol{k}\, |\boldsymbol{P}_{\mathrm{cv}}(\boldsymbol{k})|^2 \int_{-\infty}^{\infty} s^3\, \mathrm{d}s\, \exp\left(-\mathrm{i}\left[E_{\mathrm{cv}}(\boldsymbol{k}) - \hbar\omega - \mathrm{i}\Gamma\right] s\right) \tag{5.14}$$

since the first term of the approximation compensates for the unperturbed dielectric function. The s^3 term is mathematically identical to $\left(-\mathrm{i}\,\partial/\partial\left(\hbar\omega\right)\right)^3$. Thus, the change $\Delta\epsilon$ can be expressed as third derivative of the unperturbed dielectric function ϵ as[346, 347]

$$\Delta\epsilon = \frac{(\hbar\Omega)^3}{3(\hbar\omega)^2}\frac{\partial^3}{\partial(\hbar\omega)}\left[(\hbar\omega)^2\,\epsilon(\hbar\omega)\right]. \tag{5.15}$$

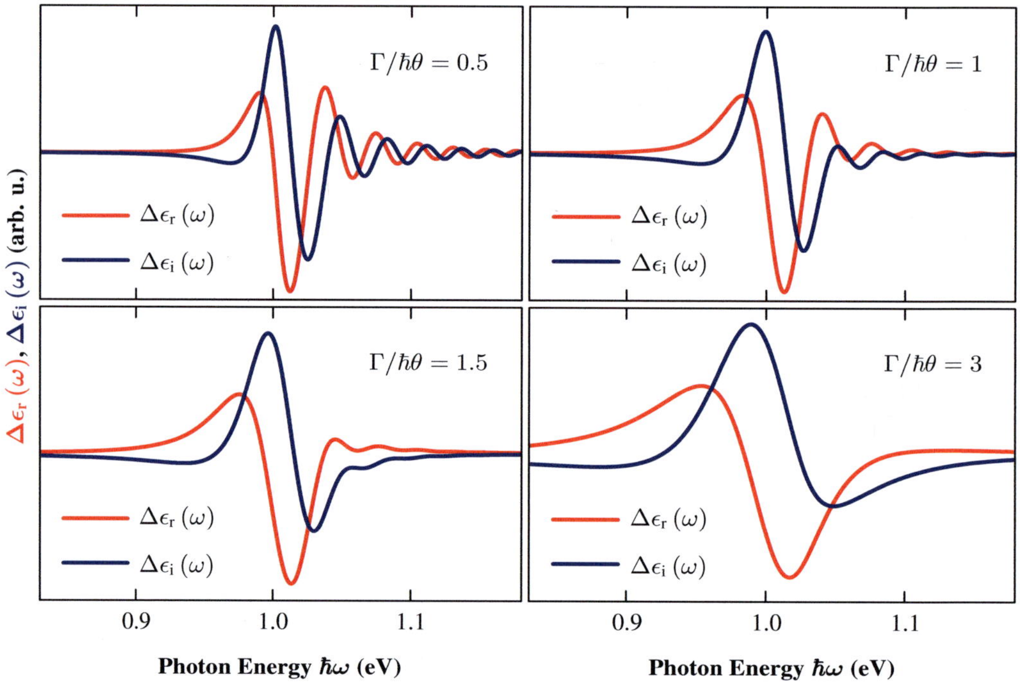

Figure 5.3: The influence of line-width broadening Γ on the change of real (red) and imaginary (blue) part of the dielectric function $\Delta\epsilon_r(\omega)$ and $\Delta\epsilon_i(\omega)$ (with $E_g = 1\,\mathrm{eV}$, $\hbar\theta = 20\,\mathrm{meV}$). With increasing Γ, the Franz–Keldysh oscillations above the bandgap energy are effectively "washed" out approaching the so-called low-field regime.

The third-derivative nature of $\Delta\epsilon$ explains the sharpness of electromodulation spectra.[346] The corresponding *Third-Derivative Functional Form* (TDFF) will be introduced in section 5.3.1. More importantly, the line shape of $\Delta\epsilon$ and, thereby, of the experimentally obtained optical spectra is *independent of the electric field* in the low-field regime.[346] Only the amplitude of the spectra scales quadratically with $\mathcal{E}$ due to $\Delta\epsilon \propto (\hbar\Omega)^3 \propto \mathcal{E}^2$ which will be further examined in chapter 6.

5.2.3 Confined States and Excitons

So far, the discussion of the field-induced implications on the dielectric function presented above completely relies on acceleration of free charge-carriers in the presence of an applied electric field. However, this mechanism is obviously not possible in case of bound states, e.g., in molecular systems, spatially-confined states, and excitons. Instead, the confinement leads to discrete and, thereby, dispersionless energies resulting in an infinite effective mass (along the confinement direction) and vanishing electro-optic energy $\hbar\theta = 0$.[356] As a consequence, the electric field does not break translational symmetry but acts as an additional linear potential on the charge-carriers which still remain confined. Then, the influence of the additional electric field is rather weak compared to the confining potential and can be modeled as small perturbation in the framework of *Stark theory*. The shift in energy ΔE induced by an external electric field $\mathcal{E}$ is described by[372, 373]

$$\Delta E\,(\mathcal{E}) = -\boldsymbol{m}_{\text{if}} \cdot \mathcal{E} - \frac{1}{2}\,(p_{\text{if}}\,\mathcal{E}) \cdot \mathcal{E} \tag{5.16}$$

with $\boldsymbol{m}_{\text{if}}$ and p_{if} being the change in dipole moment and polarizability, respectively, between initial and final state of a transition. The first (latter) term refers to the linear (quadratic) *Stark effect*. The resulting change $\Delta\alpha$ of the absorption can be obtained by expanding it as a Taylor series and inserting equation (5.16):[373–375]

$$\Delta\alpha = \frac{\partial\alpha}{\partial E}\Delta E + \frac{1}{2}\frac{\partial^2\alpha}{\partial E^2}\,(\Delta E)^2 + ... \tag{5.17}$$

$$= \underbrace{-\frac{\partial\alpha}{\partial E}\,\boldsymbol{m}_{\text{if}} \cdot \mathcal{E}}_{=0 \text{ for isotropic samples}} \underbrace{-\frac{1}{2}\frac{\partial\alpha}{\partial E}\,p_{\text{if}}\,\mathcal{E}^2}_{\text{induced dipoles}} \underbrace{+\frac{1}{2}\frac{\partial^2\alpha}{\partial E^2}\,(\boldsymbol{m}_{\text{if}} \cdot \mathcal{E})^2}_{\text{permanent dipoles}} + ... \tag{5.18}$$

The first linear contribution vanishes for isotropic samples. The second (third) term describes the influence of induced (permanent) dipoles and the associated first-(second-)derivative characteristic of the electromodulated absorption. Since excitons play the most significant role in the context of semiconducting perovskites and do not possess a permanent dipole moment, the focus of this work is on so-called *first-derivative spectroscopy*. It should be noted here that the electromodulated optical features based on the quadratic Stark effect scale quadratically with electric field $\mathcal{E}$ as in the low-field limit of free charge-carriers.

Extensive studies on excitons in the context of electromodulation have demonstrated that the modulating electric field $\mathcal{E}_{AC}$ does not only modify the resonance energy of an optical transition E_{res} but can also alter the wavefunction overlap (intensity I) and, thereby, the oscillator strength as well as the lifetime broadening Γ.

Accordingly, the change of the dielectric function $\Delta\epsilon$ is given by the sum of the respective first derivatives[356, 357, 376, 377]

$$\Delta\epsilon = \left[\left(\frac{\partial\epsilon}{\partial E_{res}}\right)\left(\frac{\partial E_{res}}{\partial\mathcal{E}_{AC}}\right) + \left(\frac{\partial\epsilon}{\partial\Gamma}\right)\left(\frac{\partial\Gamma}{\partial\mathcal{E}_{AC}}\right) + \left(\frac{\partial\epsilon}{\partial I}\right)\left(\frac{\partial I}{\partial\mathcal{E}_{AC}}\right)\right]\mathcal{E}_{AC} \tag{5.19}$$

The corresponding line shape is termed a *First-Derivative Functional Form* (FDFF) and will be introduced in section 5.3.2.

For the sake of completeness, it should be mentioned that, for sufficiently strong electric fields or small exciton binding energies, the exciton might be ionized leading to a breakdown of Stark theory. This more complex situation has been treated, e.g., by Blossey who introduced an ionization field $\mathcal{E}_I$ to combine excitonic and unbound third-derivative properties.[341, 378] Rowe and Aspnes pursued an approach including a so-called "contact" interaction with strength g, which is localized within one unit cell, resulting in an altered change of the dielectric function $\overline{\Delta\epsilon} = \Delta\epsilon / \left[1 + g\left(\epsilon - 1\right)\right]^2$.[342, 346, 379]

5.3 Evaluation and Line Shapes

The fundamental physical considerations of the electric-field influence on the dielectric function ϵ of a semiconductor in the previous chapter led to a third-derivative behavior of the change $\Delta\epsilon$ for unbound charge-carriers in the low-field limit and a first-derivative characteristic for excitons. In this section, the resulting line shapes for the evaluation of experimental spectra are presented, which are named *Third-Derivative Functional Form* (TDFF) and *First-Derivative Functional Form* (FDFF), respectively. Additionally, for the precise and facile determination of the resonance energy of an optical transition independent of such assumptions on the physical mechanism, a so-called *modulus transformation* which is based on Kramers–Kronig relations can be utilized.

5.3.1 Third-Derivative Functional Form (TDFF)

The third-derivative form of (5.15) can be explicitly calculated for parabolic bands using (3.7) and including a finite lifetime broadening Γ (by substituting $\hbar\omega - E_g$ with $\hbar\omega - E_g + i\Gamma$):[347]

$$\Delta\epsilon_i \propto \frac{(\hbar\theta)^3}{\omega^2}\frac{1}{\left(\hbar\omega - E_g + i\Gamma\right)^{5/2}} \tag{5.20}$$

This result can be generalized by a complex resonance line shape leading to *Aspnes' Third Derivative Functional Form* (TDFF) for the experimentally accessible relative change in reflectance R (or transmittance T)[347, 380]

$$\frac{\Delta R}{R}\left(\text{or } \frac{\Delta T}{T}\right) = \text{Re}\left[\frac{Ce^{\text{i}\varphi}}{(\hbar\omega - E_\text{g} + \text{i}\Gamma)^n}\right] \tag{5.21}$$

where C and φ are amplitude and phase factor which vary only slowly with ω. n equals $5/2$ for a M_0 critical point in three dimensions. The quadratic scaling of the signal with the amplitude of the electric field $\boldsymbol{\mathcal{E}}$ is implemented in the amplitude C, whereas the asymmetry of the line shape is governed by the phase factor φ. Figure 5.4 shows a TDFF line shape (with $n = 5/2$) and its strong dependence on the phase factor φ.

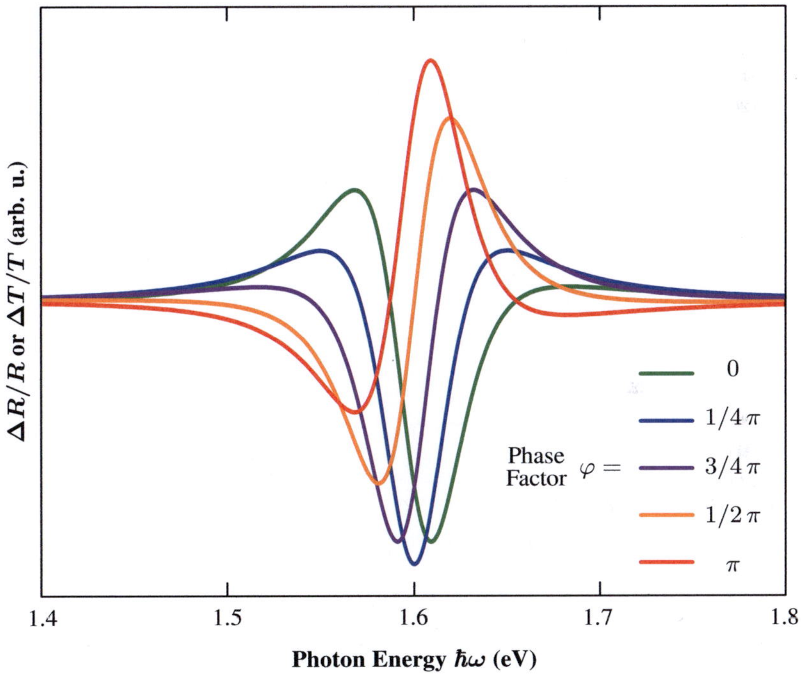

Figure 5.4: Dependence of the Third-Derivative Functional Form (for $n = 5/2$) on the phase factor φ (with $E_\text{g} = 1.6\,\text{eV}$, $\hbar\theta = 40\,\text{meV}$). Clearly, the asymmetry of the line shape is mainly governed by φ.

5.3.2 First-Derivative Functional Form (FDFF)

In the case of bound states, in particular excitons, the change of the real and imaginary part of the dielectric function, $\Delta\epsilon_r$ and $\Delta\epsilon_i$, in (5.19) can be expressed as[356, 381]

$$\Delta\epsilon_\xi = \left[A_E f_E^{(\xi)} + A_\Gamma f_\Gamma^{(\xi)} + A_I f_I^{(\xi)}\right] (I/\Gamma)\, \mathcal{E}_{\mathrm{AC}}, \qquad \xi = \{\mathrm{r}, \mathrm{i}\} \tag{5.22}$$

with

$$\begin{aligned} A_E &= (1/\Gamma)\,(\partial E_{\mathrm{res}}/\partial \mathcal{E}_{\mathrm{AC}}) & f_E^{(\xi)} &= (\partial\epsilon_\xi/\partial E_{\mathrm{res}}) \\ A_\Gamma &= (1/\Gamma)\,(\partial \Gamma/\partial \mathcal{E}_{\mathrm{AC}}) & f_\Gamma^{(\xi)} &= (\partial\epsilon_\xi/\partial \Gamma) \\ A_I &= (1/I)\,(\partial I/\partial \mathcal{E}_{\mathrm{AC}}) & f_I^{(\xi)} &= (\partial\epsilon_\xi/\partial I) \end{aligned} \tag{5.23}$$

The modulation terms $f_E^{(\xi)}$, $f_\Gamma^{(\xi)}$, $f_I^{(\xi)}$ can be calculated for a single resonance depending on the assumed broadening mechanism which will be either Lorentzian or Gaussian. The respective mathematical expressions for the assumed dielectric functions ϵ and resulting modulation contributions $f_E^{(\xi)}$, $f_\Gamma^{(\xi)}$, $f_I^{(\xi)}$ for both broadening types are shown in appendix C.1.

The treatment of individual resonances, e.g., discrete excitonic energy levels, with suitable broadening mechanism in case of FDFF line shapes apparently differs from the description of band-to-band transitions with a parabolic band dispersion of (quasi-)free electrons for TDFF line shapes.

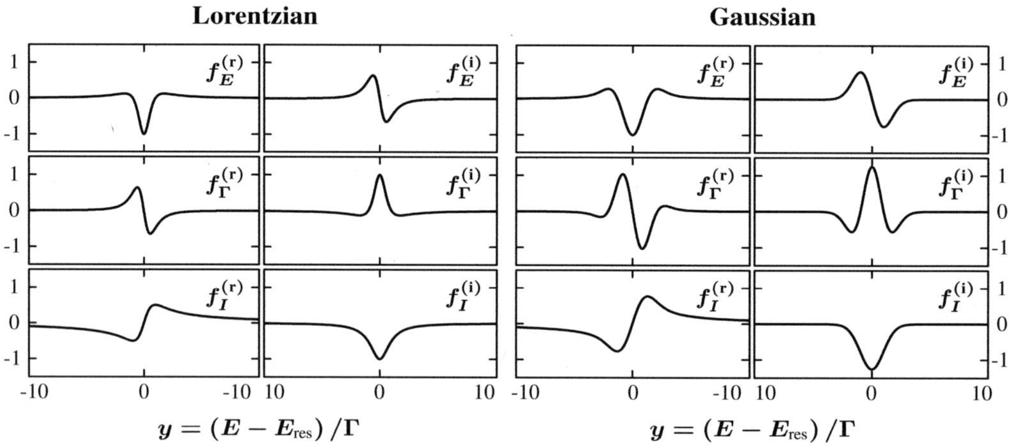

Figure 5.5: Calculated line shapes for the derivatives of the real and imaginary part of the dielectric function ϵ for (left) Lorentzian and (right) Gaussian broadening with respect to resonance energy E_{res}, broadening Γ and intensity I. Mathematical expressions can be found in appendix C.1. In case of Lorentzian broadening, $f_E^{(r)}$ and $-f_\Gamma^{(i)}$ as well as $f_E^{(i)}$ and $f_\Gamma^{(r)}$ are identical reducing the number or necessary terms for line-shape fitting. Similar is approximately true for Gaussian broadening.

Figure 5.5 displays the resulting functions $f_E^{(\xi)}$, $f_\Gamma^{(\xi)}$, $f_I^{(\xi)}$ for both broadening types. If intensity modulation terms $f_I^{(\xi)}$ can be neglected in the case of Lorentzian broadening only $f_E^{(\mathrm{r})}$ and $f_E^{(\mathrm{i})}$ remain as independent modulation factors because they are identical to $-f_\Gamma^{(\mathrm{i})}$ and $f_\Gamma^{(\mathrm{r})}$, respectively (see Fig. 5.5). The residual line shape can then be expressed analytically by (5.21) with $n = 2$.[356] For Gaussian-broadened line shapes, similar approximated expressions can be found.[356] Thus, in most cases, $f_E^{(\mathrm{r})}$ and $f_E^{(\mathrm{i})}$ are sufficient for proper fitting of experimental spectra.[382] A detailed discussion of differences of the modulation terms $f_E^{(\xi)}$, $f_\Gamma^{(\xi)}$, $f_I^{(\xi)}$ in line-shape fits of electromodulation spectra investigated in this work is presented in appendix C.2.

5.3.3 Modulus Transformation based on Kramers–Kronig Relations

An alternative method for the precise evaluation of the bandgap energy (or in general the energy of an optical resonance) was developed by Hosea based on Kramers–Kronig relations (see (3.2)).[354] The measured spectrum, here exemplary $\Delta R/R$, is treated as real part of a complex quantity $\mathcal{M}(E)$. Since $\Delta\epsilon$ and, thereby $\Delta R/R$, are causal functions, the corresponding imaginary part $\mathcal{I}(E) = \mathrm{Im}[\mathcal{M}(E)]$ can be calculated using Kramers–Kronig relations. The so-called *modulus spectrum* is then obtained by[73, 354]

$$|\mathcal{M}(E)| = \left[\left(\frac{\Delta R(E)}{R(E)}\right)^2 + \mathcal{I}^2(E)\right]^{1/2}. \tag{5.24}$$

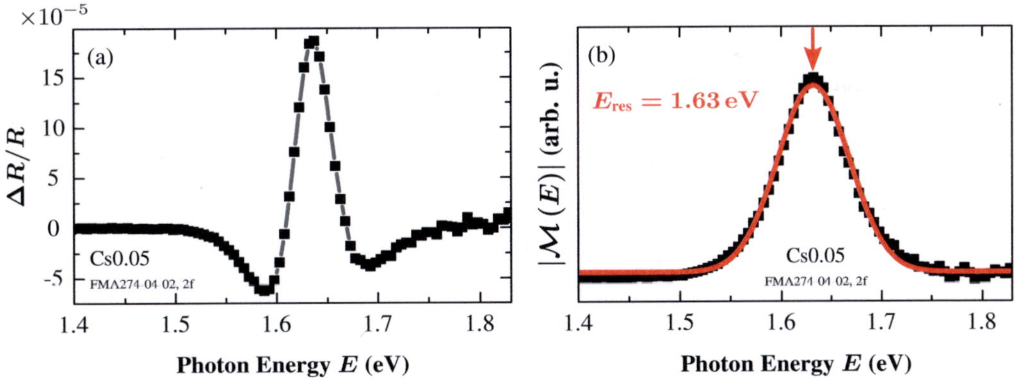

Figure 5.6: Transformation of (a) the experimentally obtained $\Delta R/R$ spectrum to (b) the peak-like so-called modulus spectrum $|\mathcal{M}(E)|$ based on Kramers–Kronig relations. The energy of the optical resonance is equivalent to the peak position and can be easily determined independent of any line-shape assumptions, e.g., by fitting with a Gaussian function (red curve). Data published in [241, 383].

The working principle behind this operation is easily demonstrated by application to the general form of a TDFF in (5.21)[73, 354]

$$|\mathcal{M}(E)| = \left| \frac{Ce^{i\varphi}}{((E - E_g) + i\Gamma)^n} \right| = \frac{C}{\left((E - E_g)^2 + \Gamma^2\right)^{n/2}}. \tag{5.25}$$

The resulting spectrum exhibits a "pseudo-Lorentzian" peak function no longer depending on the phase factor φ which strongly influences the line shape of the TDFF (illustrated in Fig. 5.4).[354] The peak position coincides with E_g and can be easily determined without the necessity of any assumptions on the critical-point type and the exponent n. Figure 5.6 presents an exemplary evaluation of a $\Delta R/R$ spectrum of a solar cell with $MAPbI_3$ absorber layer.

5.4 Summary and Comparison with Absorption Spectroscopy

Modulation spectroscopy is a powerful measurement technique for the exploration of the electronic structure at critical points in semiconductors. The fundamental principle relies on the experimental detection of small relative changes in reflectance or transmittance spectra ($\Delta R/R$ or $\Delta T/T$) with respect to a – in most cases – external perturbation, e.g., temperature, stress or, electric field. These periodic modifications affect the electronic band structure via the joint density of states and, thereby, the dielectric function and related optical properties. The underlying physical mechanisms can either conserve or break the translational symmetry of the crystal lattice which results in line shapes which can be described by either first or higher-order derivatives of the dielectric function.

Electromodulation spectroscopy is the most suitable type for the application on photovoltaic devices, even on multilayer thin-film solar cells, and is therefore utilized within this work. The periodic perturbation by an applied electric field can accelerate free charge carriers and break the translation symmetry along the field direction which allows non-vertical transitions in the band structure. The corresponding theoretical foundation has been reviewed in this chapter in the framework of Franz–Keldysh theory resulting in characteristic oscillations above the bandgap energy. In the low-field limit, the change of the dielectric function can be approximated by the third derivative of the dielectric function itself leading to a Third-Derivative Functional Form (TDFF). In case of confined and, in particular, excitonic states, the charge-carries cannot be accelerated by the electric field and lattice periodicity is maintained. For excitons, this results in a first-derivative line shape, a so-called First-Derivative Functional Form (FDFF). Both line shapes scale with the square of the amplitude of the modulating electric field which will be further discussed in the following chapter. Additionally, the spectra can be evaluated by an assumption-independent modulus transformation which is predicated on Kramers–Kronig relations.

The main benefit of modulation spectroscopy resides in the capacity to probe the electronic band structure and, in particular, to determine the energy of critical points and the respective optical transitions very precisely due to sharp resonance features in the spectra. Moreover, the measurement of relative changes of reflectance or transmittance effectively eliminates the dependency on influences such as detector response or intensity fluctuations of the illumination source. In case of solar cells, the application of an ac bias to modify the electric field in the device is straightforward and the possibility to investigate opaque samples in reflection geometry is advantageous in many cases.

However, in some instances, absorption spectroscopy can provide insights into material properties which are not accessible by electromodulation spectroscopy. The careful evaluation of absorption spectra utilizing a generalized Elliott formula or f-sum rule enables the determination of exciton binding energy E_B and bandgap energy E_g. In particular, the relationship of the bandgap energy and the resonance energy obtained from electromodulation spectroscopy requires an in-depth examination for a correct interpretation of the results which can be achieved in correlation with absorption spectroscopy. However, the acquisition of absorption data depends on the capacity of fabricating semitransparent samples, i.e. solar cells with transparent contact layers, which is not always feasible.

In this work, both spectroscopic techniques are utilized to gain deeper insights into the properties of perovskite solar cells depending on the appropriate use case: analysis of absorption spectra, e.g., for determination of exciton binding energies in different perovskite compounds (chapter 7.2), electromodulation spectroscopy, e.g., for precise studies of bandgap instabilities (chapter 9), and the combination of both to investigate the nature of fundamental optical resonances (chapter 7.1) or phase-transition suppression in multiple-cation compounds (chapter 8).

Chapter 6

Electromodulation Spectroscopy of Perovskite Solar Cells

Modulation Spectroscopy is a powerful tool for the electro-optic characterization of semiconductors in various applications ranging from bulk materials[59–66] to quantum-well structures[67–73] to solar cells.[74–76, 341, 384–387] In the latter case, in particular, electromodulation is perfectly suited since electric contacts for the application of an external electric field are already integrated in the device structure. In recent years, electroreflectance spectroscopy has been successfully applied to thin-film solar cells based on $Cu(In,Ga)Se_2$ or $Cu_2ZnSn(S,Se)_4$ absorber layer.[74, 75] Within this work, electroreflectance and electroabsorption spectroscopy are successfully applied to perovskite-based thin-film solar cells.

This chapter presents the experimental realization of electromodulation techniques (section 6.1) as well as peculiarities of the application to perovskite solar cells. By comparison with established results of $Cu(In,Ga)Se_2$- and $Cu_2ZnSn(S,Se)_4$-based solar cells, optical and electronic aspects relevant for electromodulation spectroscopy are discussed: interference-related line-shape distortions due to multilayer stacks (section 6.2.1), reflectance and transmittance configuration (section 6.2.2), influence of frequency and waveform of the modulating voltage (section 6.3), and differences in case of first and second harmonic detection (section 6.4).

In particular, the possibility of detecting reflectance and transmittance enables the investigation of both standard opaque samples with metal back contact and semitransparent solar cells for comparison with absorption results. Furthermore, the correlation of spectra detected at first and second harmonic frequency with varying amplitude of the modulating voltage represents the basis for a non-destructive determination of the built-in potential. Respective fundamental features are demonstrated utilizing simulations of already established $Cu(In,Ga)Se_2$-based solar cells.

6.1 Experimental Realization of Electromodulation Spectroscopy

The basic concept for the experimental realization of electromodulation spectroscopy can be understood based on Fig. 6.1. The light of a tunable light source consisting of a monochromator (focal length $f = 0.32\,\mathrm{m}$, 600 lines/mm) equipped with a 250 W halogen lamp is focused onto the solar-cell sample. Light from higher diffraction orders is blocked by a filter (F1) in front of the monochromator. The reflected light is imaged onto an electrically-cooled silicon photodiode using a lens system (L3 and L4). The sample is positioned on a double-rotation mount to enable a continuous and independent choice of *angle of incidence* (AOI) and *angle of detection* (AOD), both measured against the surface normal of the sample. The required electromodulation of the sample's optical properties is achieved by applying an ac bias with frequency f using a function generator. This reference signal is passed on to two lock-in amplifiers for phase-sensitive detection of the change in reflected light intensity "ΔR"[1] (the ac part of the photodiode output). The averaged reflected light intensity "R"[1] (the dc part of the detector signal) is simultaneously recorded by a multimeter. The final $\Delta R/R$ spectrum is calculated using a custom *LabVIEW* program. By using two lock-in amplifiers, the $\Delta R/R$ spectrum can be simultaneously recorded at the fundamental modulation frequency f and the second-harmonic frequency $2f$. Further details on this aspect, which is especially useful for perovskite-based solar cells, is presented in section 6.4.

The measurement setup has been extended in collaboration with Pascal Rietz within his bachelor's thesis[383] in order to simulate solar-cell working conditions. A compact solar simulator provides approximately an AM1.5 solar spectrum. In order to control the surrounding atmosphere of the solar-cell sample, a plexiglass[2] chamber has been integrated which can be flooded by compressed air, pure nitrogen or oxygen. The gas is partly guided through a gas washing bottle filled with distilled water to absorb additional humidity. The relative humidity (% RH) in the chamber can be adjusted by two flow controllers changing the ratio of gas with and without increased humidity and is monitored by a hygrometer. On the front side of the chamber, a quartz glass window is used to reduce optical losses leading to a decrease of the signal amplitude of less than 10% without any apparent distortions.

Furthermore, temperature-dependent measurements ($T = 4.2\,\mathrm{K} - 295\,\mathrm{K}$) of electroreflectance and electroabsorption spectra are possible using a liquid-helium cryostat, which is depicted in Fig. 6.2. Therefore, mirror M1 (in Fig. 6.1) is removed to guide the light of the tunable illumination source (via M2) to the sample which is mounted inside the cryostat in this case (in Fig. 6.2). Depending on the configuration of mirror M2, both reflectance (via L5) or transmittance (via M3 and L6) measurements are feasible to obtain electroreflectance or electroabsorption spectra.

[1] "R" and "ΔR" are not the actual reflectance R and its change ΔR since the measured signals are not corrected for the spectral dependence of illumination intensity and detector response or compared to a reference. However, these influences cancel out in the final $\Delta R/R$ spectrum which is one of the advantages of this measurement technique.

[2] polymethyl methacrylate (PMMA)

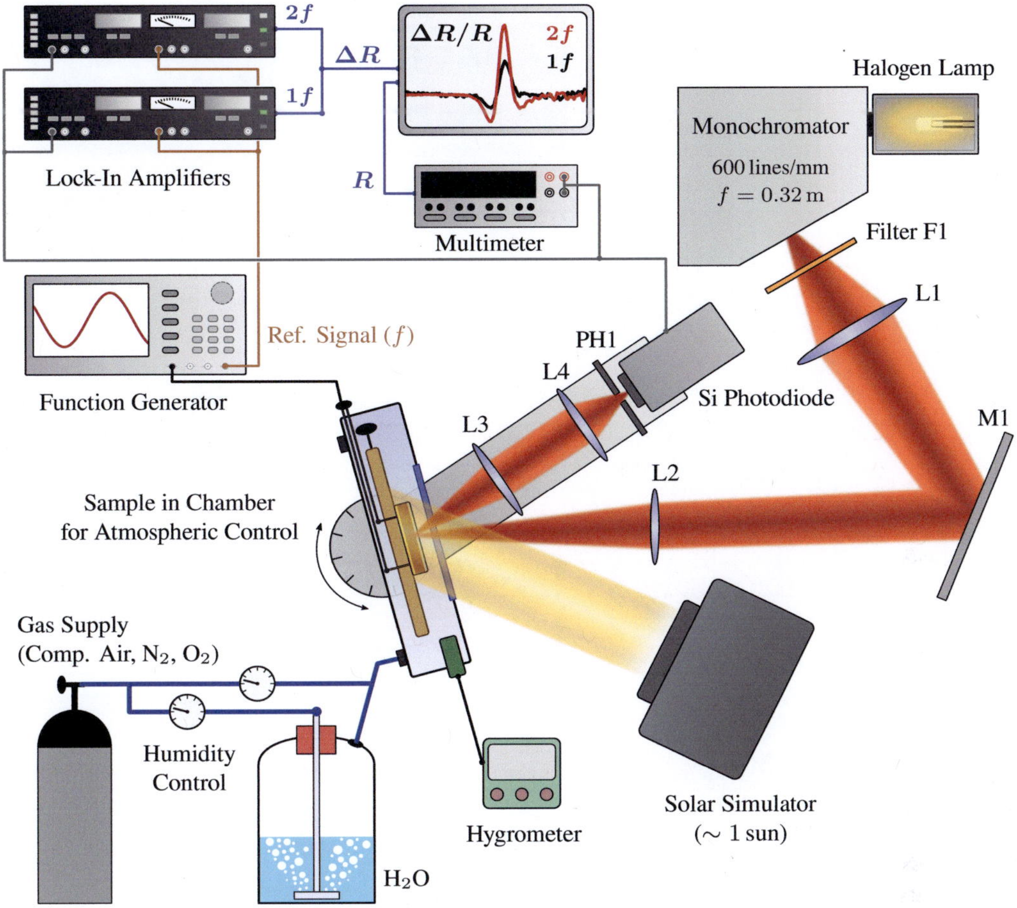

Figure 6.1: Experimental realization of room-temperature electromodulation spectroscopy. Light of a tunable light source consisting of a halogen lamp spectrum filtered by a monochromator is focused onto the sample. The reflected light is imaged onto a silicon detector and detected by a multimeter ("R"[1]). An ac bias with frequency f (and, if necessary, a dc offset) is applied to the investigated solar cell and two lock-in amplifiers are used for phase-sensitive detection ("ΔR"[1]) at first and second harmonic frequency ($1f$ and $2f$, respectively). Angle of incidence and detection on the sample can be varied continuously and independently by a double-rotation mount. In order to control the atmosphere surrounding the investigated solar cell, the chamber can be flooded with different gases (compressed air, pure N_2 or O_2) with adjusted humidity. Additionally, solar irradiation can be simulated with a solar simulator. By removal of mirror M1, the tunable light source can be used for temperature-dependent measurements (see Fig. 6.2). Additional information are presented in appendix A.3. Visualization partly based on [383].

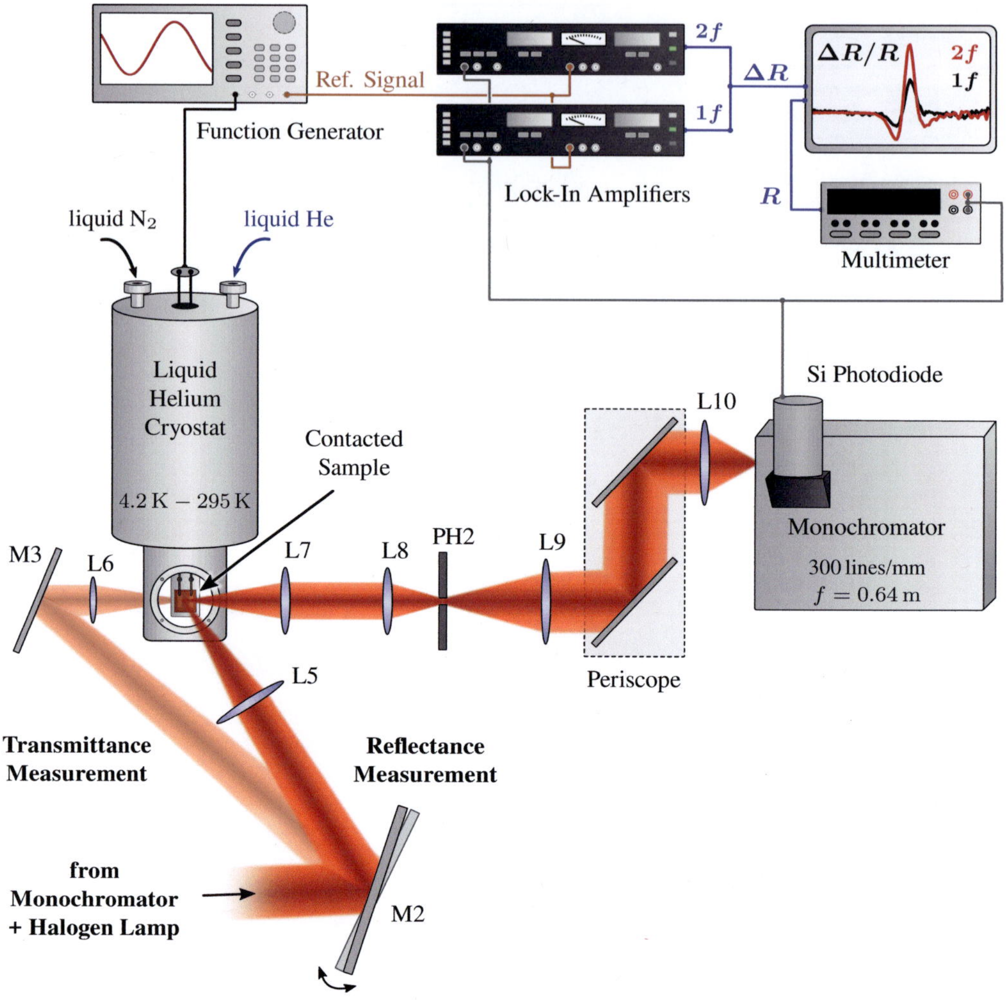

Figure 6.2: Realization of temperature-dependent electroreflectance and electroabsorption spectroscopy. Light of the same tunable light source shown in Fig. 6.1 is focused onto the solar-cell sample which is mounted inside a liquid-helium cryostat ($T = 4.2\,\mathrm{K} - 295\,\mathrm{K}$). Analogously to Fig. 6.1, an ac bias with frequency f is applied to it, and "R" and "ΔR" are detected by a multimeter and two lock-in amplifiers (at $1f$ and $2f$), respectively. An additional monochromator in the detection beam path (in front of the silicon photodiode) which is tuned synchronously to the illumination light source can be used to suppress unwanted light such as luminescence from the sample. Otherwise the silicon photodiode is placed directly behind pinhole PH2. Depending on the orientation of mirror M2, reflected or transmitted light from the sample is detected. In the latter case, absorption spectra as presented in Fig. 4.1 and chapter 4.1 can be obtained by recording a spectrum of the illumination source before or after measuring the sample's spectra. Further details can be found in appendix A.3.

Additionally, absorption spectra can be calculated similar to the setup presented in Fig. 4.1 (see chapter 4.1) by recording a spectrum of the light source before or after measuring the sample's spectrum. In the detection path, the reflected or transmitted light of the sample is imaged (via L7 and L8) onto a pinhole (PH2) to block stray light and focused onto the entry slit of a second monochromator. The latter is scanned synchronously with the illumination wavelength and can be used to suppress background light or luminescence from the sample. Alternatively, the silicon photodiode can be placed directly behind the pinhole (PH2). The sample is contacted using gold spring probes (as in the room-temperature setup) to apply the modulating ac voltage inside the cryostat. The detection of "R" ("T") and "ΔR" ("ΔT") and calculation of $\Delta R/R$ ($\Delta T/T$) are performed similarly to the room-temperature case in Fig. 6.1.

Further details about the components of the setup can be found in appendix A.3. The choice of typical measurement parameters is discussed in the following sections based on experimental tests and comparison with well-established electroreflectance spectroscopy on other thin-film solar-cell types.

6.2 Influence of Measurement Geometry on Electromodulation Spectra

In order to establish electromodulation spectroscopy on perovskite solar cells, the influences of measurement setup and parameters need to be examined carefully. For instance, the multilayer nature of thin-film solar cells (with thicknesses comparable to wavelengths in the visible spectrum) can give rise to interference effects. In fact, this is one of the major challenges for electromodulation spectroscopy of $Cu_2ZnSn(S,Se)_4$- and $Cu(In,Ga)Se_2$-based solar cells causing severe line-shape distortions of the measured spectra.[74,75] Furthermore, differences between measurements in reflection and transmission are evaluated since both configurations can enable useful applications, e.g., comparison with absorption results or investigation of standard solar-cell samples with metal back contact.

Major parts of initial testing and verification of electromodulation spectroscopy on perovskite solar cells presented in this and the following chapter have been achieved in collaboration with Ina Kelz within her master's thesis.[388] Temperature-dependent electroreflectance and electroabsorption results have been obtained together with Alice Magin within her master's thesis.[389]

6.2.1 Evaluation of Interference-Related Line-Shape Distortions in Electroreflectance Spectra

In order to assess the influence of interference effects on the line shape of electromodulation spectra, measurements with varying angle of incidence (AOI) and angle of detection (AOD) have been performed. Figure 6.3 (a) shows normalized $\Delta R/R$ spectra measured in specular reflection geometry, i.e., AOI $=$ AOD, for different angles of incidence ranging from $15°$ to $40°$. Clearly, no distortions of the spectra are observable, thereby, excluding any influence of interference effects on the evaluation. This observation is rather surprising because of the multi-layer structure of these thin-film solar cells and contrasts electroreflectance spectra of, e.g., Cu(In,Ga)Se_2-based devices. Nevertheless, the occurrence of interferences is influenced by different parameters such as layer thicknesses, refractive index contrast and interface roughness. Optical simulations indicate that there are indeed interference-related effects present in reflectance and transmittance spectra of perovskite solar cells but, fortunately, they do not significantly affect the investigated energy region close to the bandgap energy.[390–392]

For further evidence, additional measurements of *diffuse electroreflectance* have been carried out. This technique has been developed by Christoph Krämmer *et al.* specially for the suppression of interference-related distortions in electroreflectance spectra of $Cu_2ZnSn(S,Se)_4$- and Cu(In,Ga)Se_2-based thin-film solar cells.[75] By detecting diffusely scattered light, the effective angular interval of the collected light is much larger compared to the narrow angle in specular reflection.[75] Thereby, interference-related oscillations in $\Delta R/R$ spectra are washed out enabling an evaluation with standard Third-Derivative Functional Form (TDFF) fits.[75] See [75] for further information.

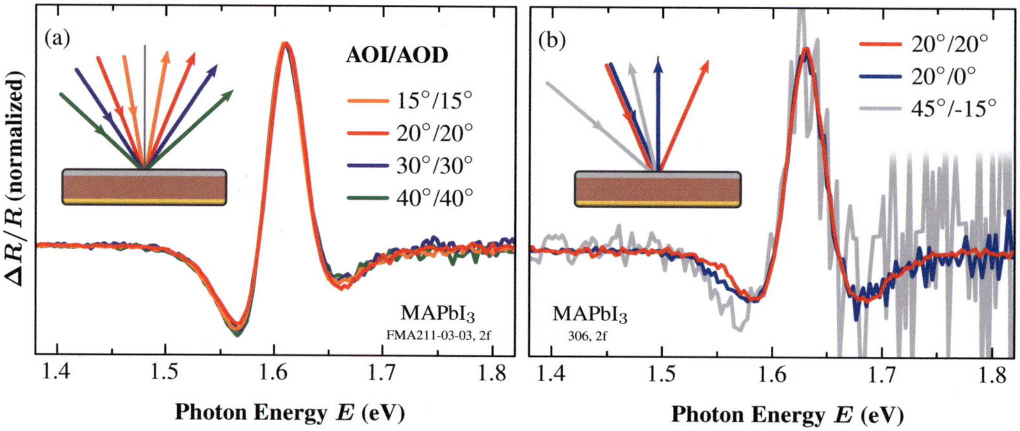

Figure 6.3: Dependence of $\Delta R/R$ spectra on angle of incidence (AOI) and angle of detection (AOD) in (a) specular and (b) diffuse (blue and grey curves) reflection geometry. In all cases no distortions of electroreflectance spectra due to interference effects are visible. Data in (b) published in [388].

In case of perovskite-based solar cells, no changes in the spectra (blue and gray curve in Fig. 6.3 (b)) compared to specular reflectance (red curve) are visible except for a strong reduction of *signal-to-noise ratio* due to a decrease in detected light intensity.

In conclusion, opposed to $Cu_2ZnSn(S,Se)_4$- and $Cu(In,Ga)Se_2$-based solar cells, interference effects do not lead to any observable line-shape distortions in electromodulation spectra of perovskite-based thin-film solar cells. Therefore, all measurements are carried out in specular reflection geometry (mainly at AOI = AOD = 15°) due to the best signal-to-noise ratio.

6.2.2 Reflection and Transmission Configuration

Because perovskite solar cells can be easily fabricated in a semitransparent design using transparent electrodes, they can be used in various ways – not only in terms of practical applications, but also with regard to spectroscopy. Therefore, the comparability of electromodulation results obtained in reflection and transmission geometry is an important question.

Figure 6.4 (a) presents normalized experimental $\Delta T/T$ (red curve) and $\Delta R/R$ (blue curve) spectra of a semitransparent solar cell with $MAPbI_3$ absorber layer and transparent FTO[3] and IZO[4] electrodes. Both line shapes exhibit a small difference concerning the relative amplitudes of their two side minima. However, this does not necessarily mean that the extracted resonance energy is inconsistent but rather that the mixing of real and imaginary part of the change $\Delta\epsilon$ of the dielectric function into $\Delta R/R$ and $\Delta T/T$ is different as described by the Seraphin coefficients in equations (5.1) and (5.2) (also illustrated by the phase factor φ in Fig. 5.4). The equal resonance energies are confirmed by nearly identical peak positions of Gaussian fits to the transformed modulus spectra $|\mathcal{M}(E)|$ in Fig. 6.4 (b).

An additional comparison of extracted resonance energies E_{res} from temperature-dependent $\Delta T/T$ electroabsorption (red circles) and $\Delta R/R$ electroreflectance (blue triangles) spectra at $T = 10\,\text{K} - 295\,\text{K}$ is shown in Fig. 6.5. Overall, both measurements yield similar results but with a small but still striking divergence in the orthorhombic crystal phase below 150 K. Although the reason for this clearly needs further investigation the deviations are small compared to other relevant energy differences, e.g., between resonance energy and bandgap energy E_g (black squares) extracted from Elliott fits to absorption spectra. This comparison of E_{res} and E_g is subject of a more elaborate discussion in chapter 7.1.

In summary, both reflection and transmission geometry yield similar results with only minor deviations enabling the evaluation of the spectra which suit the respective use case best. The comparison with absorption spectra in chapter 7 requires the use of semitransparent samples favoring electroabsorption spectroscopy, while phase-transition studies in chapter 8 can be performed on standard opaque solar cells.

[3] **F**luorine-doped **T**in **O**xide
[4] **I**ndium-doped **Z**inc **O**xide

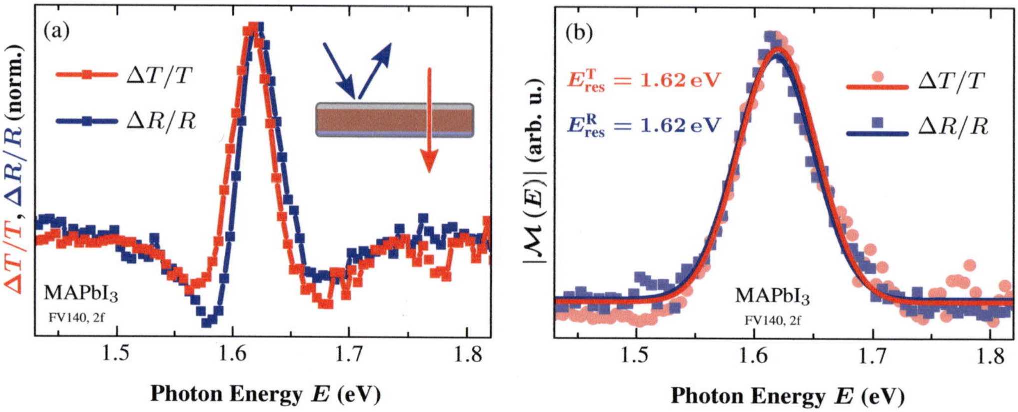

Figure 6.4: (a) Comparison of normalized $\Delta T/T$ (red curve) and $\Delta R/R$ (blue curve) spectra measured on a semitransparent solar cell with $MAPbI_3$ absorber layer and transparent FTO[3] and IZO[4] contact layers at room temperature. Despite small differences in relative intensities of the side minima, no significant deviation of the peak position in the transformed modulus spectra in (b) is observable resulting in very similar resonance energies $E_{\mathrm{res}}^{\mathrm{T/R}}$ (determined by Gaussian fits depicted as red and blue curves). Data published in [388].

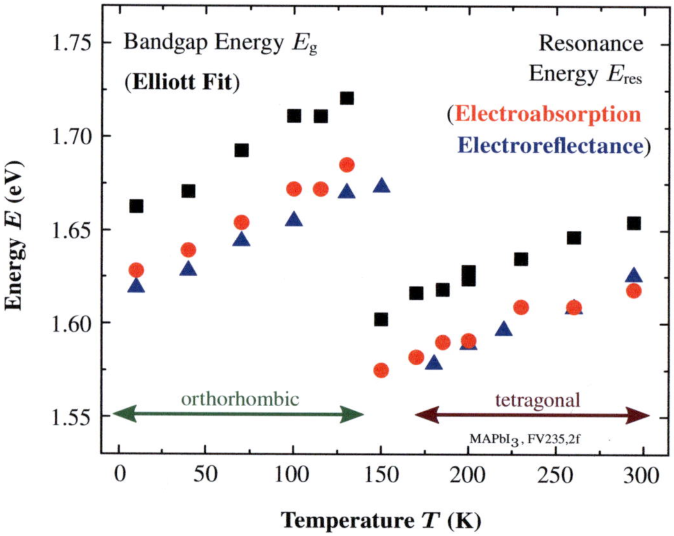

Figure 6.5: Comparison of resonance energies obtained from transformed modulus spectra of measured electroabsorption (red circles) and electroreflectance (blue triangles) spectra on a semitransparent solar cell with $MAPbI_3$ absorber at $T = 10\,\mathrm{K} - 295\,\mathrm{K}$. Both yield similar results with only minor deviations especially in the low-temperature orthorhombic crystal phase. For comparison with bandgap energy E_{g} extracted from Elliott fits to standard absorption spectra, see chapter 7.1. © 2018 IEEE. Reprinted, with permission, from [358].

6.3 Identification of Suitable Modulation Voltages

In addition to optical aspects, it is important for reliable measurements to achieve a modulation of the electric field in the absorber layer while preventing a current flow through the device since this could lead to distortions in the recorded spectra. In order to inhibit any current injection in the perovskite, an additional insulating layer can be integrated in the layer stack of the sample, e.g., a PMMA[5] interlayer, which is typically utilized to measure electromodulation spectra of perovskites.[341, 374, 393] Figure 6.6 (a) presents first electroreflectance measurements which have been performed on modified devices (red curve) with an approximately 230 nm-thick PMMA[5] layer instead of the hole-transporting spiro-OMeTAD layer (see inset for scanning electron micrograph of sample cross section). In this case, comparably high amplitudes of the modulation voltage can be applied, here $V_{ac} = 5\,V$, compared to an unmodified standard device (black curve) with $V_{ac} = 0.8\,V$.

However, concerning practical applications, measurements on standard solar cells without the necessity of modifying their layer stack are strongly desirable. Moreover, any modifications of

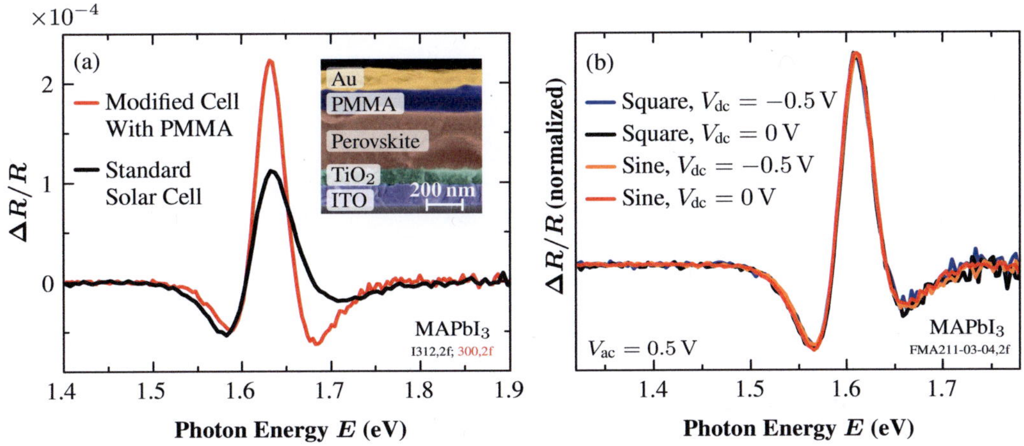

Figure 6.6: (a) Electroreflectance spectrum for a $MAPbI_3$-based solar cell in standard configuration (black curve) and with insulating PMMA[5] layer (red curve) instead of spiro-OMeTAD as hole-transport layer. With additional insulating layer higher amplitudes of the modulation voltage are possible and needed (here $V_{ac} = 5\,V$ compared to $V_{ac} = 0.8\,V$ for the standard cell). Inset: Scanning electron micrograph of modified device architecture with PMMA[5] layer (dark blue). (b) Normalized $\Delta R/R$ spectra of a standard solar cell for sine-wave (orange and red curve) and square-wave (blue and black) voltage in reverse bias condition (orange and blue) and symmetric around $V_{dc} = 0\,V$ (black and red). All spectra do not show any differences indicating that no line-shape distortions due to current injection occur. Data in (a) published in [388].

[5] polymethyl methacrylate (PMMA)

the layered solar-cell structure (especially of layers in direct contact with the perovskite absorber) can change the properties of the investigated system (see, e.g., the examination of bandgap instabilities in chapter 9) and should therefore be avoided. For Cu(In,Ga)Se_2-based solar cells, electromodulation spectroscopy of complete standard solar cells is realized by applying a voltage in reverse bias.[74,75] Accordingly, square-wave (as used for Cu(In,Ga)Se_2) and sine-wave modulation voltages in reverse bias (blue and orange curves, with dc offset $V_{dc} = -0.5\,\mathrm{V}$) are compared to symmetric wave-forms around $V_{dc} = 0\,\mathrm{V}$ (black and red curves) for standard perovskite solar cells in Fig. 6.6 (b). Due to the missing current-blocking insulating layer, much smaller modulation amplitudes have to be used to prevent significant current flow and sample damage, here $V_{ac} = 0.5\,\mathrm{V}$. Still, well-defined electroreflectance spectra can be obtained (see Fig. 6.6 (a) and (b)). The normalized $\Delta R/R$ spectra do not exhibit any apparent differences in all four cases including reverse bias and symmetric wave-forms around $V_{dc} = 0\,\mathrm{V}$. This observation is confirmed by studies of Li *et al.* who partially observed deviations in their results for dc voltages above $V_{dc} \approx 1\,\mathrm{V}$ or below $V_{dc} \approx -0.6\,\mathrm{V}$ (at $V_{ac} = 1\,\mathrm{V}$).[129] Additionally, current measurements of the investigated devices reveal maximum values of $\sim 0.3\,\mathrm{mA}$ at $V_{ac} = 1\,\mathrm{V}$ (and $V_{dc} = 0\,\mathrm{V}$) which are comparable to reference measurements on Cu(In,Ga)Se_2 solar cells for which electroreflectance spectroscopy can be considered as well-established technique.[388]

In conclusion, electromodulation spectroscopy is successfully applied to perovskite solar cells with standard layer stack (without additional modifications needed) using either sine-wave or square-wave voltages with up to $V_{ac} = 1\,\mathrm{V}$ and $V_{dc} = 0\,\mathrm{V}$ or under reverse bias conditions. Within this work, sine-wave voltages are used since they do not contain higher-order frequency components (in contrast to square waveforms). If not otherwise stated, $V_{ac} = 0.5\,\mathrm{V} - 1\,\mathrm{V}$ and $V_{dc} = 0\,\mathrm{V}$ are applied. The dependence of electromodulation results on the applied modulating frequency f is part of the following section in the context of detection at fundamental and second harmonic frequency.

6.4 Detection at Fundamental and Second-Harmonic Frequency

In case of electromodulation applied to a sample by an external electric field (induced by an applied ac bias) with frequency f, one would intuitively expect to detect changes in the optical properties of the sample at the same frequency f. However, for a modulating electric field $\mathcal{E}$ consisting of a constant offset $\mathcal{E}_0$ and an oscillating ac part described by

$$\mathcal{E} = \mathcal{E}_0 + \mathcal{E}_{ac}\cos(\Omega t) \qquad \text{with } \Omega = 2\pi f, \tag{6.1}$$

the resulting change in the dielectric function relevant for modulation spectroscopy can be expressed as[341]

$$\Delta\epsilon \propto \mathcal{E}^2 = \left(2\,\mathcal{E}_{ac}\mathcal{E}_0\cos(\Omega\,t) + \frac{1}{2}\,\mathcal{E}_{ac}^2\,[1 + \cos(2\,\Omega\,t)] + \mathcal{E}_0^2\right). \tag{6.2}$$

This is justified because $\Delta\epsilon$ scales with the square of the electric field in the low-field limit of free charge carriers as well as for excitonic states. As a result, (6.2) contains both a contribution varying with applied modulation frequency Ω (or f, in red) and an additional term oscillating with the second-harmonic frequency $2\,\Omega$ ($2f$, in blue). It should be noted that the constant offset $\mathcal{E}_0$ can consist of an externally applied field $\mathcal{E}_{dc}$ and an internal electric field $\mathcal{E}_{int}$.

Experimentally, spectra can be obtained at both the fundamental frequency f and the second-harmonic frequency $2f$ which is shown in Fig. 6.7 (a) and (b), respectively. While the line shape of the normalized $\Delta R/R$ spectra is independent of the modulation frequency f at $2f$ detection, severe distortions occur in the $1f$ case, in particular with decreasing frequency f. Therefore, in most cases, stable electromodulation spectra with good signal-to-noise ratio are recorded at $2f$ detection and higher frequencies (In the following, $f = 990\,\text{Hz}$ is used unless otherwise mentioned.).[341, 393] In contrast to this, electromodulation spectra of other established thin-film solar cells such as Cu(In,Ga)Se_2 are commonly obtained at the fundamental frequency. The reason for this exceptional behavior at $1f$ detection could be rather unstable internal fields affecting only $1f$ spectra since $\mathcal{E}_0$ and $\mathcal{E}_{int}$ only enter the contribution varying with Ω (in red) but not $2\,\Omega$ (in blue) in equation (6.2). These unstable internal fields could be related, e.g., to ion migration processes at corresponding millisecond to second time scales.[129, 272–274]

For deeper insights into electromodulation spectra at $1f$ and $2f$ detection, measurements with varying amplitude of the modulating ac voltage V_{ac} at constant $V_{dc} = 0\,\text{V}$ have been performed (Fig. 6.8 (a) and (b)). Afterwards, the amplitudes A_{ER} of the resonance feature in the electroreflectance spectra have been extracted by Gaussian fits to the transformed $|\mathcal{M}\,(E)|$

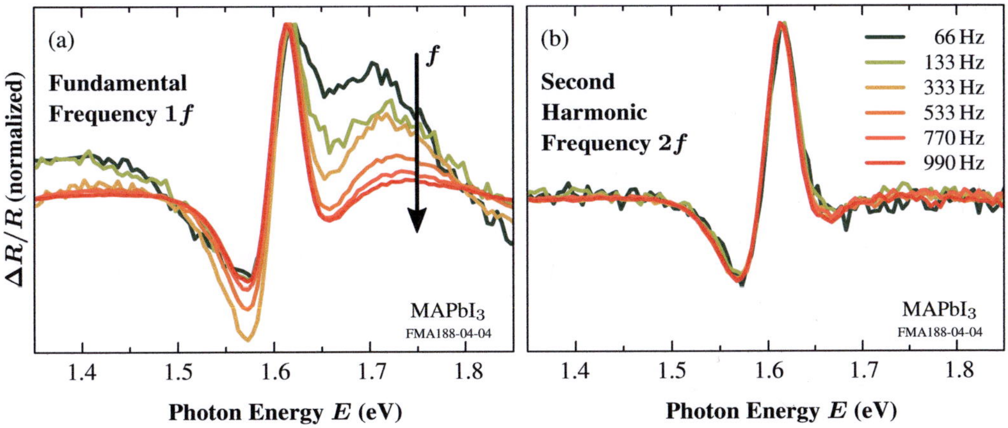

Figure 6.7: Dependence of electroreflectance spectra of a perovskite solar cell with $MAPbI_3$ absorber layer on modulation frequency f in a range of 66 Hz – 990 Hz detected at (a) fundamental frequency $1f$ and (b) second harmonic frequency $2f$. In case of $1f$ detection, line-shape distortions are observable, in particular, at low modulation frequencies while at $2f$ detection all normalized $\Delta R/R$ spectra show an identical line shape. Best signal quality and signal-to-noise ratio are achieved for second-harmonic detection at $f = 990\,\text{Hz}$. Data published in [388].

modulus spectra and are shown in Fig. 6.8 (c) and (d) for $1f$ and $2f$ detection, respectively (The error bars represent the fit uncertainties only and are negligible in case of $2f$ detection. However, the total uncertainties are presumably larger especially for the rather noisy signals at $1f$ detection.). While at $1f$ detection, the dependence of the amplitude A_{ER} on modulation amplitude V_{ac} can be reasonably well fitted by a linear relationship, A_{ER} scales with the square of V_{ac} in case of $2f$ detection. This is also confirmed by a slope of 2.06 ± 0.03 in a double logarithmic plot (see inset of Fig. 6.8 (d)) and similar results reported by, e.g., Wu *et al.*[393] and Ziffer *et al.*[341] These dependencies match the predictions in (6.2) for the naive assumption that the $\Delta R/R$ amplitude scales similarly with V_{ac} as $\Delta\epsilon$ increases with $\mathcal{E}_{ac}$.

In order to further substantiate these findings, simulations of electroreflectance spectra of well-established Cu(In,Ga)Se_2-based thin-film solar cells have been conducted since required material parameters are already better known in this system compared to perovskite solar cells. The calculations of the electronic structure and electric fields in the device for a discretized applied sine-wave voltage are performed by the one-dimensional simulation program *SCAPS* developed at the *University of Gent*.[394–396] Based on these electric field distributions, optical simulations are performed using a transfer-matrix method approach initially developed by Christian Huber and Christoph Krämmer.[74, 397, 398] Finally, a *Fast-Fourier Transform* (FFT) is utilized to obtain electroreflectance spectra at $1f$ and $2f$ detection frequencies from the calculated time-domain data. Initial simulation results (as well as the experimental data in Fig. 6.8) have been obtained in collaboration with Alice Magin within her master's thesis[389] and have been further improved together with Ellen Förstner. Further details on the simulation procedure can be found in appendix D and the above mentioned references.

Figure 6.9 presents simulation results on a thin-film solar cell with Cu(In,Ga)Se_2 absorber layer similarly visualized as the experimental results on $MAPbI_3$-based solar cells in Fig. 6.8. It should be noted that the simulations are conducted for reverse bias conditions with $V_{dc} = -900\,\mathrm{mV}$ in accordance with typical measurement conditions for these solar cells. The amplitude of the $\Delta R/R$ spectra at $1f$ detection is apparently larger than for $2f$ detection which matches one's intuitive expectation but is opposed to the results for $MAPbI_3$-based solar cells. The oscillations at higher photon energies which appear for increasing voltage amplitudes can be attributed to Franz–Keldysh oscillations. Figures 6.9 (c) and (d) clearly verify the experimental observations of a linear trend of the amplitude A_{ER} of $\Delta R/R$ spectra on modulation amplitude V_{ac} in case of $1f$ detection and the quadratic behavior in the $2f$ case (notice a slope m close to 2 in the double logarithmic plot in the inset).

In addition to the intuitive assumption for the dependency of $\Delta R/R$ amplitude A_{ER} on ac modulating voltage V_{ac} derived from the relationship between $\Delta\epsilon$ and ac modulating electric field $\mathcal{E}_{ac}$, more detailed simulations could provide deeper insights into these relations and underlying principles. In particular, the spatial distribution of the electric field in the device and its influence on the resulting spectra need further investigation. Ultimately, a detailed understanding of these effects can provide an electro-optic measurement technique for non-destructive investigations of internal fields in thin-film solar cells by comparison of voltage-dependent scaling of electroreflectance spectra at $1f$ and $2f$ detection (see $\mathcal{E}_0$ in both terms in (6.2)).

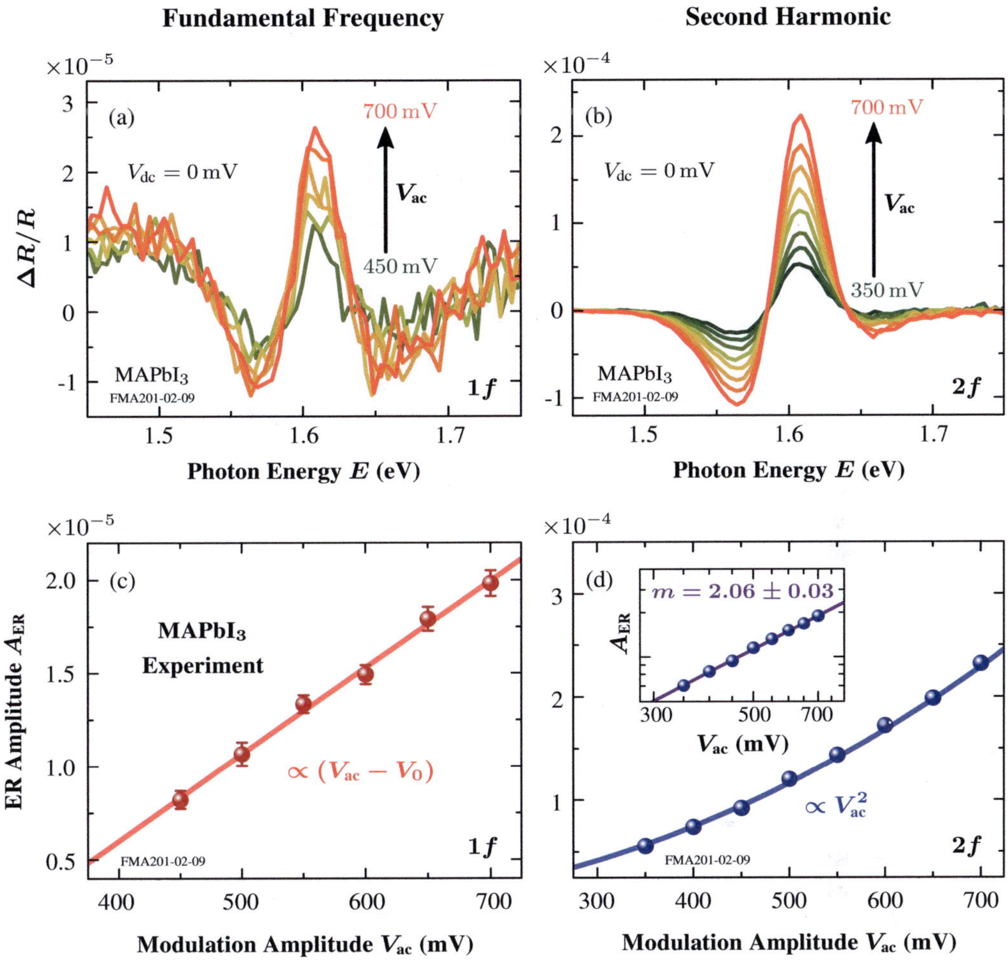

Figure 6.8: Experimental electroreflectance spectra for varying amplitude of modulation voltage $V_{ac} = 350\,mV - 700\,mV$ of a $MAPbI_3$-based solar cell in case of (a) $1f$ and (b) $2f$ detection. The corresponding amplitudes A_{ER} have been extracted from Gaussian fits to the transformed $|\mathcal{M}(E)|$ modulus spectra, again for (c) $1f$ and (d) $2f$ detection (The error bars represent the fit uncertainties only and are negligible in case of $2f$ detection. However, the total uncertainties are presumably larger especially for the rather noisy signals at $1f$ detection.). A_{ER} depends linearly on V_{ac} for $1f$ detection and scales with the square of V_{ac} in the $2f$ case. The latter is confirmed by a slope m of 2.06 ± 0.03 in the double logarithmic plot (see inset). These observations seem to confirm the naive expectations from (6.2) assuming a similar relationship between amplitude A_{ER} and V_{ac} compared to $\Delta\epsilon$ and $\mathcal{E}_{ac}$. Data published in [389].

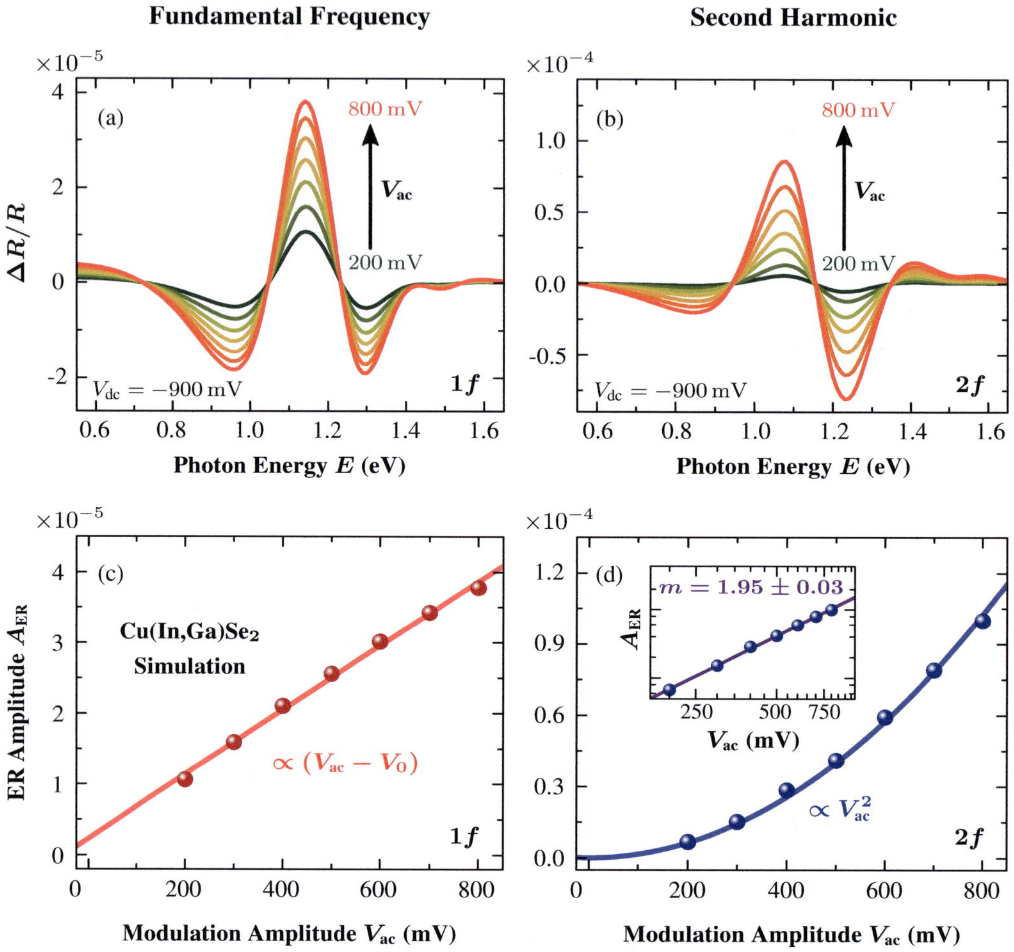

Figure 6.9: Simulated electroreflectance spectra for varying amplitude of modulation voltage $V_{ac} = 200\,\mathrm{mV} - 800\,\mathrm{mV}$ at $V_{dc} = -900\,\mathrm{mV}$ (reverse bias condition) of a Cu(In,Ga)Se_2-based solar cell in case of (a) $1f$ and (b) $2f$ detection. $\Delta R/R$ amplitudes are larger in case of $1f$ detection as intuitively expected and Franz–Keldysh oscillations arise at higher modulation voltages. The linear scaling of the electroreflectance amplitude A_{ER} with increasing ac modulation amplitude V_{ac} at $1f$ detection in (c) as well as the quadratic dependence in the $2f$ case in (d) (see also the double logarithmic plot in the inset with a slope m close to 2) confirm experimentally observed trends. Data obtained in collaboration with Ellen Förstner.

6.5 Summary

In this chapter, the experimental realization of electromodulation spectroscopy used within this work is presented. This includes the possibility of additional illumination with an approximately AM1.5 solar spectrum, and controlled surrounding gas atmosphere and relative humidity to simulate solar-cell working conditions. Furthermore, temperature-dependent ($T = 4.2\,\mathrm{K} - 295\,\mathrm{K}$) measurements in reflection (electroreflectance) or transmission (electroabsorption) are feasible. The comparability of the extracted resonance energies for both measurement configurations is demonstrated for the whole temperature range investigated showing only minor deviations. Consequently, both electroreflectance and electroabsorption spectroscopy can be utilized depending on the respective use case, e.g., comparison with absorption spectra for semitransparent solar cells (see chapter 7) or investigation of standard opaque devices with metal back contact (see chapter 8).

The influence of interference effects and different modulation voltages is experimentally examined and compared to results of well-established electroreflectance spectroscopy on Cu(In,Ga)Se_2- and Cu_2ZnSn(S,Se)$_4$-based thin-film solar cells. In contrast to these, electromodulation spectra of perovskite solar cells do not show any dependence of obtained line shapes on the angle of incidence and detection – neither in specular nor in diffuse reflection – proving the absence of interference-related distortions of the spectra. While Cu(In,Ga)Se_2 and Cu_2ZnSn(S,Se)$_4$ solar cells are standardly measured using reverse bias conditions to prevent a significant current flow through the device, perovskite solar cells yield identical line shapes for both reverse bias and symmetric voltages around an offset $V_{dc} = 0\,\mathrm{V}$ for comparably low ac voltage amplitudes $V_{ac} = 0.5\,\mathrm{V} - 1.0\,\mathrm{V}$.

An exceptional characteristic of thin-film solar cells with perovskite absorber layer is the strongly improved robustness of electromodulation spectra detected at the second-harmonic frequency $2f$ compared to the fundamental frequency f of the applied modulating ac bias (see also chapter 9.5.1). In particular at low frequencies, severe line-shape distortions occur in spectra at $1f$ detection which are not present in the $2f$ case. Furthermore, the expected linear dependence of the $\Delta R/R$ amplitude on the amplitude V_{ac} of the modulating ac bias at $1f$ detection and quadratic relation for the $2f$ case are verified by experimental results on $MAPbI_3$-based devices as well as numerical simulations of well-established Cu(In,Ga)Se_2 solar cells. Additional more elaborate simulations could provide deeper insights into the impact of the electric field and its spatial distribution on electroreflectance spectra and, thereby, offer a non-destructive, electro-optic measurement technique to study internal electric fields in thin-film solar cells.

After establishing electromodulation spectroscopy on complete perovskite solar cells, it is applied in the following chapters together with absorption spectroscopy.

Chapter 7

Excitonic Nature of Optical Transitions in Perovskites

The role of excitons in semiconductors, especially in photovoltaic applications, is an important research question to be addressed since it strongly influences the operation of optoelectronic devices. Therefore, the binding energy E_B of excitons is a significant measure or "figure of merit" to quantify these effects. In inorganic semiconductors, E_B is typically in the order of some meV (e.g., 4.9 meV in GaAs[343] and 14.7 meV in Si[349]) and, thereby, small compared to thermal energy at room temperature $k_B T \approx 25\,\mathrm{meV}$. Consequently, bound electron–hole pairs can be easily dissociated and move freely, and the solar-cell operation conditions are governed by free electrons and holes.[1,399] In the other extreme, organic semiconductors typically exhibit exciton binding energies in the range of 100 meV up to 1 eV.[196,197] Therefore, appropriate device architectures are needed in organic photovoltaics to enable efficient diffusion and dissociation of excitons at suitable interfaces in the device.[1,399]

Furthermore, a clear understanding of excitonic effects is also essential for the correct interpretation of experimental results in electro-optic material characterization. For instance, the commonly used technique for determination of the bandgap energy E_g in semiconductors based on linear extrapolation of a so-called Tauc plot[400–402] is simply not applicable in case of significant excitonic effects.[20,81] This is because the underlying assumption of a $(\alpha\hbar\omega) \propto (\hbar\omega - E_g)^{1/2}$ dependence of the absorption coefficient α is no longer valid (see also chapter 3.1).

Therefore, the importance of excitons for both the correct interpretation of optical spectra of organic–inorganic perovskites and the performance of corresponding optoelectronic devices, especially perovskite-based solar cells, will be diligently examined in this chapter. In particular, the nature of fundamental transitions in electromodulation spectroscopy and the consequences for correct evaluation are studied in detail for the mostly used model system $MAPbI_3$ (section 7.1). Additionally, absorption spectroscopy provides a suitable tool for the assessment of influences of compositional variations of the perovskite absorber on the exciton binding energy. Here, both the modification of the iodine–bromine ratio utilized for bandgap tuning and the incorporation of additional cations such as cesium are examined (section 7.2). A discussion of implications of excitonic effects on the absorption and charge-carrier extraction in perovskite solar cells complements this chapter (section 7.3).

7.1 Excitonic Origin of Fundamental Transitions in Electroabsoption Spectra of Perovskite Solar Cells

Electromodulation spectroscopy provides an effective instrument for the exploration of fundamental optical transitions in the band structure of semiconductors and is a precise method for determination of the corresponding resonance energies as pointed out in detail in previous chapters. In order to access its full potential on perovskites, the correlation with other characterization techniques offers deeper insights into the origin of optical transitions. In this context, the *generalized Elliott fit* evaluation of absorption spectra (chapter 4) represents a straightforward tool for comparison due to its distinction into contributions of discrete excitonic states and ionization continuum above the bandgap energy. Moreover, absorption spectra can be measured simultaneously with electroabsorption spectra ensuring precise results. Additionally, a detailed analysis of experimentally obtained electromodulation spectra based on theoretical models for the observed line shapes further supplements the full picture. Both approaches have been pursued and are presented in the following sections.

Main parts of the measurements in this section were achieved in collaboration with Alice Magin within her master's thesis[389] and are published in [76, 358].

7.1.1 Correlation of Absorption and Electroabsorption Spectroscopy

In order to reveal deeper insights about fundamental transitions, absorption and electroabsorption spectra of semitransparent perovskite solar cells with $MAPbI_3$ absorber layer (see chapter 2.2.2 for details) have been measured simultaneously at temperatures from $T = 10\,\mathrm{K}$ up to room temperature. Figure 7.1 (a) and (b) present obtained experimental results at $T = 10\,\mathrm{K}$. Due to strongly reduced broadening at cryogenic temperatures, sharp features and, in particular, a pronounced excitonic peak are apparent in the absorption data. In order to facilitate best possible comparison with electroabsorption results, the absorption spectrum in Fig. 7.1 (a) is analyzed using the generalized Elliott formula (introduced in chapter 4.2) and, thereby, decomposed into contributions of discrete excitonic peaks (green curve) and excitonic ionization continuum (blue). The fit reproduces the experimental data very well and yields a bandgap energy of 1.66 eV which is depicted by a dashed blue line. The direct comparison of absorption and $\Delta T/T$ spectra demonstrates that the distinct resonance feature in the electroabsorption results in Fig. 7.1 (b) is definitely below the bandgap energy E_g. This clearly points towards an excitonic origin of the fundamental optical transition rather than non-excitonic band-to-band transitions.

This conclusion of an excitonic resonance is further substantiated by applying the modulus transformation introduced in section 5.3.3 to the experimental $\Delta T/T$ data. The resulting $|\mathcal{M}(E)|$ modulus spectrum is shown in Fig. 7.1 (c) and offers a more appropriate comparison since the energy position of the peak-like line shape directly represents the energy of the optical resonance as discussed in section 5.3.3. The energy position of the transformed modulus peak corresponds well to the contribution of discrete excitonic peaks (green curve in (a)).

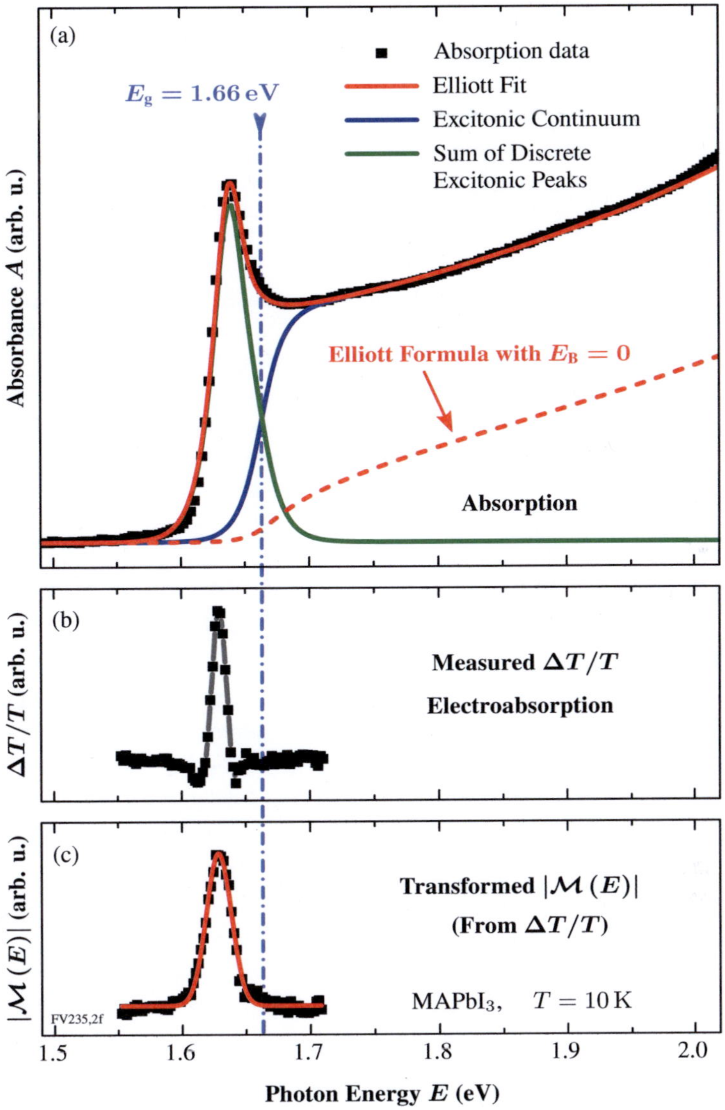

Figure 7.1: Comparison of (a) absorption, (b) electroabsorption $\Delta T/T$, and (c) transformed $|\mathcal{M}(E)|$ modulus spectrum of a perovskite solar cell with $MAPbI_3$ absorber layer at $T = 10\,\mathrm{K}$. The characteristic resonance features of both the measured $\Delta T/T$ and the transformed $|\mathcal{M}(E)|$ spectrum are clearly below the bandgap energy $E_g = 1.66\,\mathrm{eV}$ determined by a generalized Elliott fit to the absorption data in (a). The red curve in (c) visualizes a Gaussian fit to the experimental data and corresponds well to the discrete excitonic contribution (green curve) of the Elliott formula in (a) proving the excitonic nature of the fundamental optical transition in $MAPbI_3$ probed by electroabsorption spectroscopy. © 2018 IEEE. Reprinted, with permission, from [358].

The small deviations in the order of 10 meV might be caused by simplifying assumptions used in the evaluation, e.g., state-independent broadening, and treatment of all excitonic transitions as one effective resonance.[76] However, the clear agreement in energy position of the fundamental optical resonance probed by electroabsorption spectroscopy with the contribution of discrete excitonic transitions in absorption spectra proves its excitonic nature arising from bound electron–hole pairs instead of ionized excitons of the excitonic continuum above E_g.

These findings hold true not only for $T = 10\,\mathrm{K}$ but even up to room temperature since Fig. 7.1 serves only as exemplary representation of the whole investigated temperature range. Figure 7.2 (a) shows the respective energies of the bandgap E_g (black squares) obtained by Elliott fits (see Fig. 7.1 (a)) and the peak positions of the transformed modulus spectra (red circles) determined by Gaussian fits (see red curve in Fig. 7.1 (c)) to the experimental data for $T = 10\,\mathrm{K} - 295\,\mathrm{K}$. Filled and semi-filled symbols denote different samples verifying reproducibility of the results. Both bandgap energy E_g and modulus peak position exhibit an overall shift towards higher energies for increasing temperature which is typical for lead-based perovskites due to their specific electronic structure (see section 8.1). This overall blueshift is interrupted at $T_p \approx 160\,\mathrm{K}$ by a jump of about 100 meV towards lower energies which is a clear indication of the orthorhombic-to-tetragonal phase transition. Furthermore, the coexistence of both phases in the temperature range close to the phase transition can lead to a variation of the observed T_p values. Detailed investigations of crystal phase transitions in perovskites are presented in chapter 8.

Still, here, only the relative energy shift between bandgap energy E_g and modulus peak position is of interest which stays approximately constant over the whole temperature range investigated. This emphasizes that the evaluation and the corresponding results presented above are consistent even up to room temperature and further support the relevance of excitonic effects on optical transitions in perovskites.

Figure 7.2 (b) displays the temperature-dependent exciton binding energy E_B obtained as fit parameter in the generalized Elliott formula. The E_B values depend on the crystal phase and are around 26 meV for the low-temperature orthorhombic phase and 19 meV for the room-temperature tetragonal phase. Despite small deviations (as discussed above), E_B is similar to the offset of bandgap energy and fundamental excitonic resonance (from modulus peak position) in Fig. 7.2 (a) as expected from theory.

In conclusion, the overall picture of combined temperature-dependent absorption and electroabsorption results clearly reveals the excitonic nature of the fundamental transition in $MAPbI_3$ arising from the discrete contribution of excitons in the absorption spectrum rather than unbound electron–hole pairs in the ionization continuum. Furthermore, the relevance of excitonic effects in perovskites is shown – even up to room temperature.

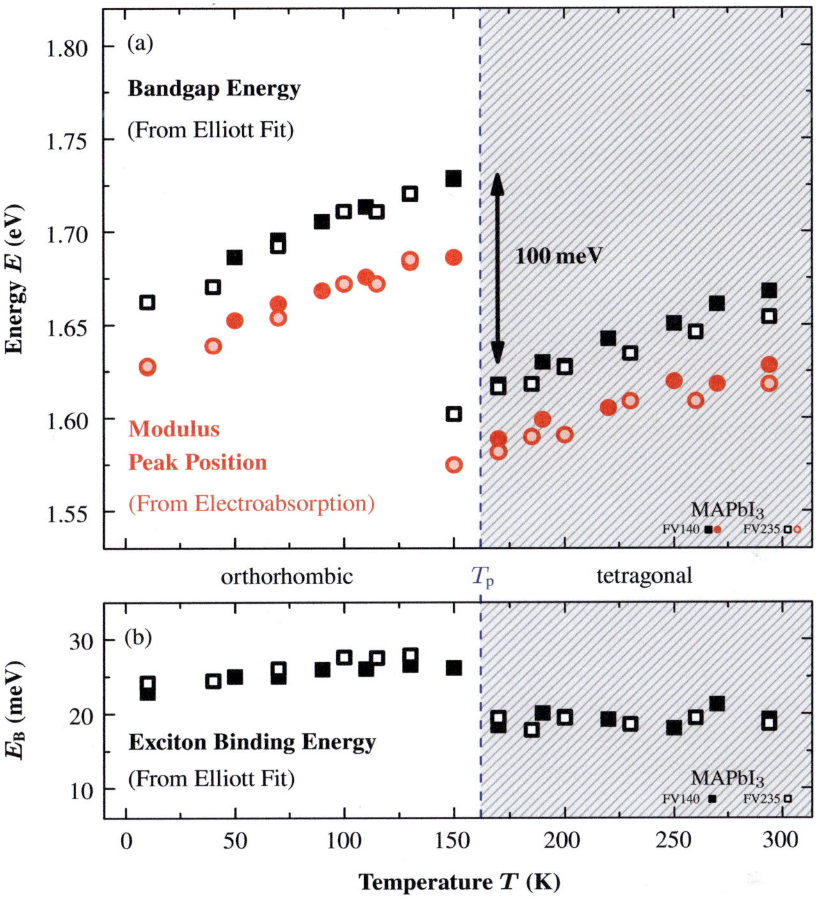

Figure 7.2: (a) Bandgap energy E_g (black squares) obtained from generalized Elliott fits of absorption spectra and energy of modulus peak position (red circles) from transformed electroabsorption spectra (as determined in Fig. 7.1) for temperatures $T = 10\,\text{K} - 295\,\text{K}$ (Filled and semi-filled symbols denote different samples verifying reproducibility of the results.). The overall blue shift of E_g and modulus peak position is typical for lead-based perovskites and the jump of about 100 meV at $T_p \approx 160\,\text{K}$ is related to the orthorhombic-to-tetragonal phase transition. The coexistence of both phases for $T \approx T_p$ can lead to a variation of observed T_p values (see chapter 8 for further discussion). The constant offset of bandgap energy and modulus peak position demonstrates the reliable evaluation and consistent results proving the importance of excitonic effects on fundamental transitions in perovskites solar cells. (b) Exciton binding energies E_B obtained by Elliot fits as in (a). The values are around 26 meV and 19 meV for the low-temperature orthorhombic and room-temperature tetragonal phases, respectively, and, thereby, in the range of the offset between E_g and the modulus peak position as expected. Previous version published in [76], further data can be found in [358].

7.1.2 Excitonic Line Shapes of Electroabsorption Spectra in Perovskites

While the correlation with absorption spectroscopy is a powerful tool, electroabsorption spectra itself can provide indications of excitonic influences by comparison of measured line shapes with theoretical predictions. An in-depth review of expected line shapes in electromodulation spectroscopy for unbound and confined states has been presented in chapter 5. In the case of band-to-band transitions of freely movable charge-carriers, a third-derivative form is predicted (in the low field limit). Confined states and, in particular, excitonic transitions result in first-derivative spectra.

Figure 7.3 depicts the numerically calculated first, second, and third derivative of a measured absorption spectrum at $T = 190\,\mathrm{K}$ and a comparison with the corresponding electroabsoption result. The best qualitative agreement of the $\Delta T/T$ spectrum with the first derivative can be considered as first hint of an excitonic line shape.

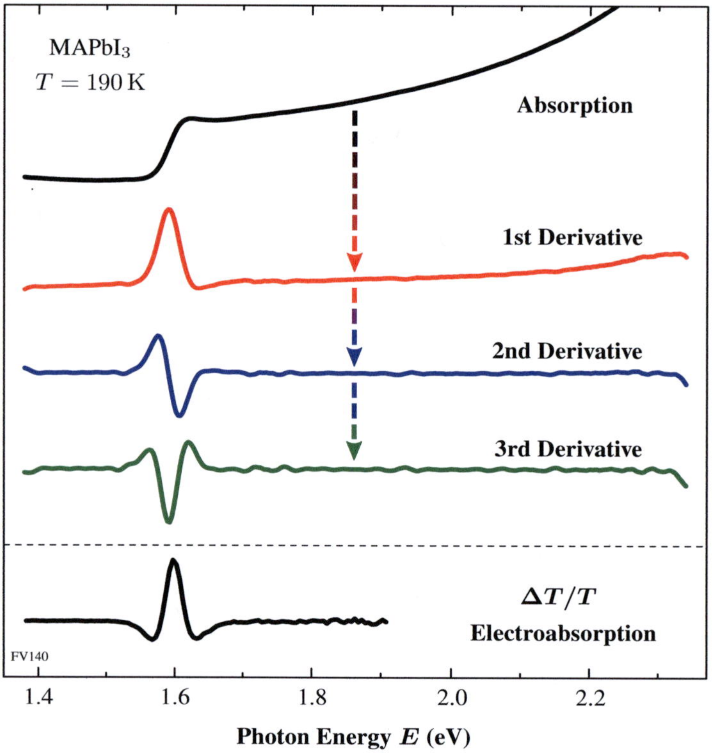

Figure 7.3: Comparison of a $\Delta T/T$ electroabsorption spectrum with the numerically calculated first, second and third derivative of the measured absorption spectrum of MAPbI$_3$ at $T = 190\,\mathrm{K}$. The best qualitative agreement with the first derivative indicates an excitonic nature of the optical resonance in electroabsorption spectra. Reprinted from [76], with the permission of AIP Publishing.

However, since additional parameters such as phase factors might influence the electromodulated spectra, only direct line-shape fits of the appropriate theoretical models introduced in chapter 5.3 can provide convincing evidence. The Third-Derivative Functional Form (TDFF) represents the predicted line shape for band-to-band transitions of (quasi-)free electrons in the low-field limit. On the other hand, a First-Derivative Functional Form (FDFF) is expected in case of excitons which possess discrete energy levels (Details on the different contributions to the FDFF line shape with respect to energy, broadening, or intensity modulation can be found in appendix C.).

The direct comparison of FDFF and TDFF line-shape fits is preceded by a determination of the suitable broadening type of the First-Derivative Functional Form. In chapter 5.3.2, the two commonly used choices have been introduced, which are Gaussian and Lorentzian broadening. Figure 7.4 displays an exemplary comparison of both broadening types, here at $T = 190\,\mathrm{K}$, either as direct least-squares fit of the respective FDFF functions (as introduced in chapter 5.3.2) to the electroabsorption spectra in (b) or as fit of Gaussian or Lorentzian functions to the transformed modulus spectrum in (c).

In both cases, the Gaussian-broadened alternative results in a better reproduction of the measured data. This is also reflected by the so-called *coefficient of determination* R^2 in Fig. 7.4 (a) which

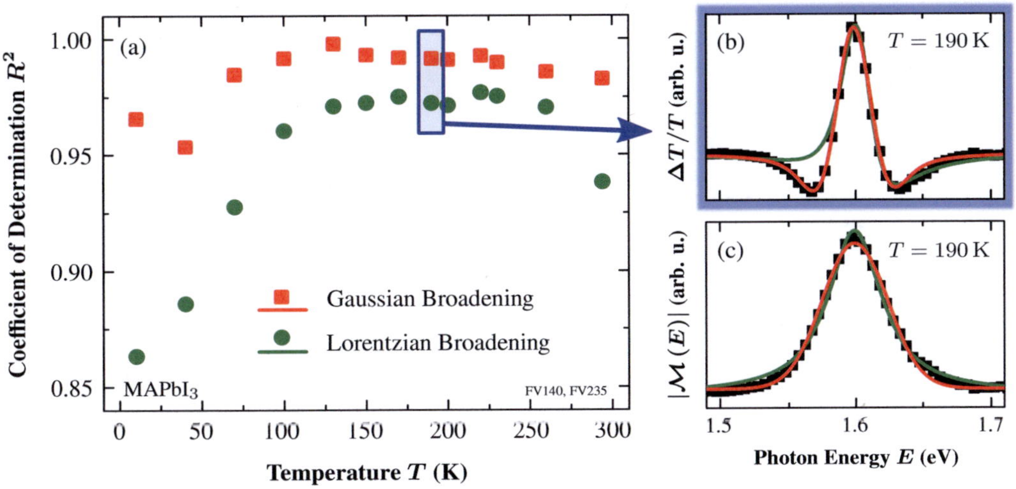

Figure 7.4: Investigation of broadening effects in $MAPbI_3$ by comparison of Gaussian (red squares and curve) and Lorentzian (green circles and dashed green curve) broadening in electroabsorption spectra. (a) Coefficient of determination R^2 for the respective FDFF fits to experimental $\Delta T/T$ electroabsorption spectra for $T = 10\,\mathrm{K} - 295\,\mathrm{K}$. The values for Gaussian-broadening FDFF fits being closer to unity indicate a better agreement with the measured data. (b) Exemplary illustration of both line-shape fits at $T = 190\,\mathrm{K}$ and (c) the corresponding fits of Gaussian and Lorentzian peak functions to the transformed $|\mathcal{M}(E)|$ spectrum. Original measured spectra obtained in collaboration with Alice Magin.[389]

describes the goodness of the fits for Gaussian- and Lorentzian-broadened FDFF line shapes, respectively, to the $\Delta T/T$ electroabsorption spectrum (see exemplary spectrum in Fig. 7.4 (b)). The clear trend towards values closer to unity for the Gaussian broadening over the whole temperature range investigated represents a better agreement with the measured data and indicates inhomogeneous broadening. Still, the underlying mechanisms, e.g., spatial variations of the resonance energy, could be in principle more complex and might not be perfectly described by a Gaussian function.[344] Nevertheless, one would not expect a completely identical resonance energy of the entire (polycrystalline) material which would result in homogeneous, Lorentzian-shaped broadening.[344] This is in accordance with the observed inferior suitability of the Lorentzian-broadened FDFF.

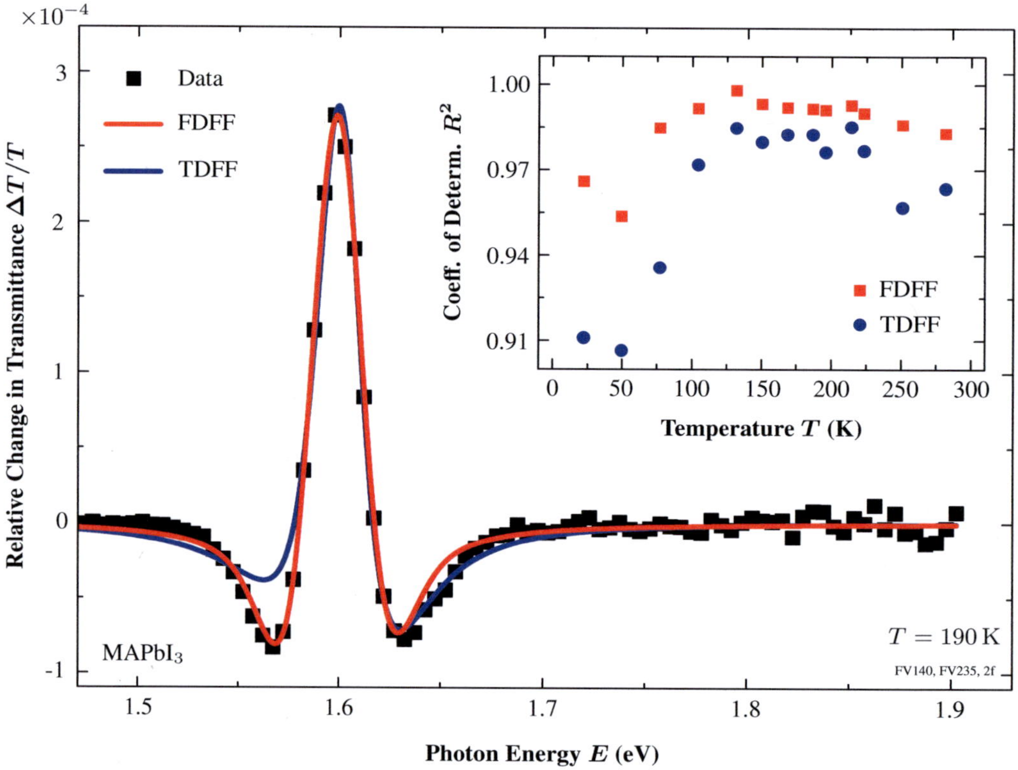

Figure 7.5: Comparison of excitonic FDFF (red curve) and non-excitonic TDFF (blue curve, expected for band-to-band transitions) least-squares fits to an exemplary experimental $\Delta T/T$ electroabsorption spectrum at $T = 190\,\mathrm{K}$. In particular, the relative intensities of the two side minima are better described by the FDFF fit which indicates an excitonic origin of the optical transition. Inset: The coefficient of determination R^2 for both fits. The conclusion of an excitonic transition is consistent for the whole investigated temperature range $T = 10\,\mathrm{K} - 295\,\mathrm{K}$. Original measured spectra obtained in collaboration with Alice Magin.[389] Exemplary spectra also published in [76, 389].

Equipped with this prerequisite, one can compare least-squares fits to experimental electro-absorption spectra by utilizing TDFF and Gaussian-broadened FDFF line shapes representing band-to-band and excitonic transitions, respectively. An exemplary comparison is shown in Fig. 7.5. While both line shapes provide a reasonable agreement with the experimental data, the excitonic FDFF line shape apparently reproduces the spectrum better and is, therefore, the more appropriate theoretical model. The deviations of the TDFF fit become especially apparent at the relative intensities of the two side minima which cannot be reproduced simultaneously. This comparative analysis has been performed for the whole temperature range measured from $T = 10\,\mathrm{K}$ up to room temperature. The resulting coefficients of determination R^2 for both line-shape fits are presented in the inset of Fig. 7.5. The goodness of the fits is a little bit reduced for low temperatures which still needs further investigation. In any case, the overall trend clearly demonstrates that the excitonic FDFF fit is unambiguously the better suited line shape. Therefore, the conclusion of an excitonic origin of the fundamental optical transitions in $MAPbI_3$ probed by electromodulation spectroscopy is further substantiated by this evaluation.

In summary, both an in-depth line-shape analysis of electromodulated $\Delta T/T$ spectra and the correlation with absorption spectra provide a clear picture of fundamental optical transitions in $MAPbI_3$: These resonances which are probed by electromodulation spectroscopy are clearly excitonic arising from discrete states of bound electron–hole pairs rather than unbound electrons and holes in the ionization continuum. In particular, it has to be emphasized that this conclusion is found to be true for the whole temperature range investigated from $T = 10\,\mathrm{K}$ up to even room temperature. These findings imply that resonance energies extracted from electromodulation spectra are not identical to the bandgap energy E_g as it would be the case in semiconductors with negligible excitonic effects, e.g., as in $Cu(In,Ga)Se_2$,[74,75] but are shifted by about the exciton binding energy towards lower values. Furthermore, the electro-optic properties of perovskites and their devices are influenced by excitons. Since the exciton binding energy quantifies the extent and, thus, relevance of excitonic effects, it will be further analysed for different perovskite compositions in the following section.

7.2 Exciton Binding Energy in Organic–Inorganic Perovskites

In order to compare the impact of excitons in different materials and, in particular, varying perovskite composition, the exciton binding energy serves as a useful measure. The exciton binding energy in organic–inorganic perovskites is still heavily discussed in the scientific community and a wide range of values for E_B has been reported. Therefore, in the following section, major, but by no means exhaustive, examples of determination methods and resulting E_B values are reviewed. Thereafter, the impact on E_B of different compositional modifications of the model system $MAPbI_3$ at both the halogen and the organic-cation site in the crystal structure (see chapter 2.1.3) is studied by absorption spectroscopy. Results of these investigations are partly published in [195].

7.2.1 Review of Published Values for the Exciton Binding Energy in Organic–Inorganic Perovskites

Different published experimental values for E_B and respective utilized methods are summarized in Tab. 7.1. In addition, discussions of experimental investigations and theoretical predictions can be found, e.g., in [21,403] and [88,95,97,404,405], respectively. Thus, relevant determination methods are only briefly highlighted here.

Years before organic–inorganic perovskites have attracted significant attention as well-suited absorber material in photovoltaics, Hirasawa *et al.* and Tanaka *et al.* used magnetoabsorption spectroscopy to determine the excition binding energy E_B of $MAPbI_3$.[339,340] By measuring the diamagnetic shift of the exciton absorption peak in magnetic fields up to $B = 40\,\mathrm{T}$ at temperature $T = 4.2\,\mathrm{K}$ they obtained an exciton binding energy E_B of 37 meV[339] and 50 meV,[340] respectively, for $MAPbI_3$ and 76 meV for $MAPbBr_3$.[340] Later, Lin *et al.* questioned their estimate for the dielectric permittivity of $\epsilon = 6.5$ and recalculated E_B to $\sim 2\,\mathrm{meV}$ (using $\epsilon \sim 70$).[406]

	E_B [meV]		Method	Ref.
$MAPbI_3$	2		Recalc. of [339, 340]	[406]
	7.4	(tetr.)	Electroabsorption	[341]
	16 ± 2	(orth.)	High-B Magnetoabs.	[338, 407, 408]
	a few	(tetrag.)		
	19 ± 3, 32 ± 5		Photoluminescence	[409, 410]
	34 ± 3	(orth.)	Absorption Fits	[79, 351, 352]
	$23 - 29$	(tetr.)		
	45	(orth.)	T-Dep. Absorption	[411]
	37, 50	(orth.)	Magnetoabsorption	[339, 340]
$MAPbI_{3-x}Cl_x$	14	(orth.)	High-B Magnetoabs.	[407]
	10	(tetr.)		
	55 ± 20	(tetr.)	T-Dep. Absorption	[412]
$MAPbBr_3$	25	(orth.)	High-B Magnetoabs.	[407]
	60 ± 3		Absorption Fits	[352]
	76	(orth.)	Magnetoabsorption	[340]

Table 7.1: Summary of published values for the exciton binding energy E_B in organic–inorganic perovskites and utilized determination methods. If no crystal phase (orthorhombic or tetragonal) is specified, data from cryogenic and room temperature have been combined in calculation or fitting procedure.

Revised magnetoabsorption measurements with increased magnetic fields up to $B = 65\,\mathrm{T}$ (even exceeding 130 T for short-pulsed experiments) allow for more precise studies of $1s$ and $2s$ excitonic resonances as well as additional Landau levels.[338, 407, 408] Exciton binding energies and effective masses can be extracted by fitting the dependence of the respective energy levels on the applied magnetic field (a so-called "Landau fan" diagram) using elaborate numerical calculations which promise more accurate results. However, the influence of strong magnetic fields $B > 50\,\mathrm{T}$ on E_B still remains somewhat unclear. The obtained values are in general smaller compared to the initial studies[338, 407, 408] and similar for both thin films and single crystals indicating the intrinsic nature of the obtained values.[408]

Detailed analysis of temperature-dependent photoluminescence and absorption spectra has led to numerous E_B estimates. Some of them rely on the determination of the temperature-dependent behavior (e.g., of line-width broadening or intensity) of the excitonic absorption or photoluminescence peak and a subsequent Arrhenius evaluation[409, 410, 412] yielding E_B values of, e.g., 19 meV[409] or 32 meV.[410] However, these results should be considered as rather rough estimates due to large uncertainties in determination and interpretation. Least-squares fitting of generalized versions of the Elliott formula or similar advanced evaluation methods as introduced in chapter 4 yield E_B values of $\sim 34\,\mathrm{meV}$[352] and $23\,\mathrm{meV} - 29\,\mathrm{meV}$[79, 351, 352] for the orthorhombic and tetragonal phases, respectively, in $MAPbI_3$ as well as 60 meV for $MAPbBr_3$.[352] Compared to Arrhenius evaluations, these direct fits of absorption spectra can be expected to yield more reliable results – at least for sufficiently small temperature-induced broadening.

In summary, there is still a wide range of published exciton binding energies for $MAPbI_3$ from as low as 2 meV up to 50 meV. However, the vast majority of E_B values is in the order of thermal energy at room temperature with an overall trend to smaller values for the high-temperature tetragonal phase.

In the following, the exciton binding energy E_B of various perovskite compounds including mixed-halide and mixed-cation compositions is studied using temperature-dependent absorption spectra and generalized Elliott fits (introduced in chapter 4).

7.2.2 Variation of Iodine–Bromine Ratio Used for Bandgap Tuning and its Impact on the Exciton Binding Energy

Bandgap engineering is one of the key issues for the application of perovskites as large-bandgap top cell in tandem configurations (see chapters 2.1.3 and 2.3). The partial replacement of iodine by bromine offers strong tunability of the perovskite's bandgap energy in a range of about $1.6\,\mathrm{eV} - 2.3\,\mathrm{eV}$ which is illustrated by experimental absorption spectra in Fig. 7.6 at $T \approx 275\,\mathrm{K}$ close to room temperature. However, the spectra do not only demonstrate the wide-range tunability but also show an increasingly pronounced excitonic peak close to the band edge for higher bromine content x. This is an indication for an increased exciton binding energy in bromine-rich compounds which could hamper charge-carrier separation and collection and, thereby, decrease the performance of corresponding photovoltaic devices.

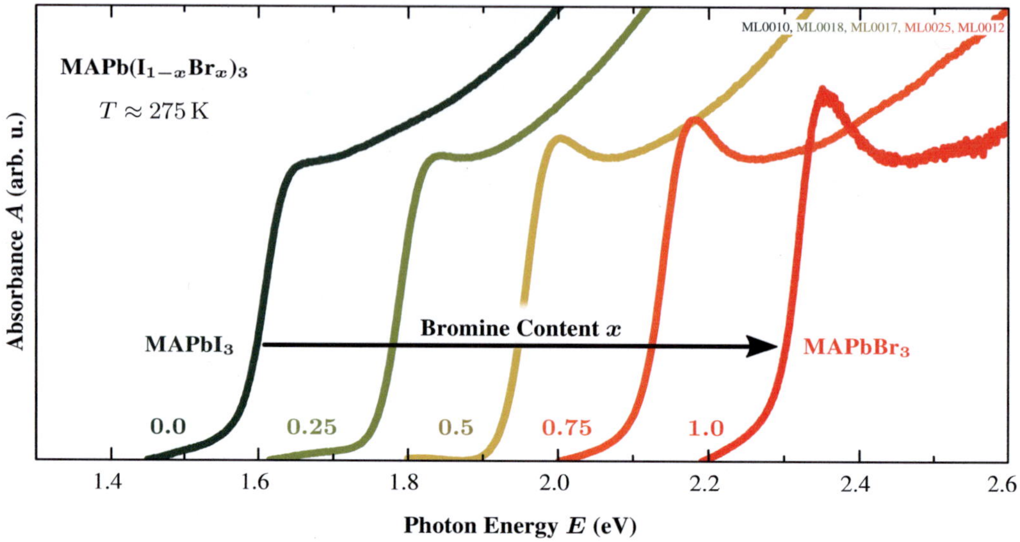

Figure 7.6: Absorption spectra of $MAPb(I_{1-x}Br_x)_3$ thin films on glass with varying bromine content $0 < x < 1$ illustrating the wide-range tunability of the bandgap energy from about 1.6 eV up to 2.3 eV. But, in addition, the increasingly distinct excitonic peak close to the band edge for increasing bromine content x indicates an enhanced exciton binding energy. It should be noted here, that any instabilities of the perovskite for intermediate bromine ratios x as reported by Hoke *et al.*[335,413] do not preclude stable and reliable measurements as can be seen from these spectra.

Here, it should be noted that Hoke *et al.* observed bandgap instabilities in photoluminescence spectra related to a photo-induced halide segregation of the perovskite for intermediate iodine–bromine ratios.[335,413] Similar effects will be reviewed and elucidated extensively in chapter 9. Still, they do not prevent the measurement of stable and reliable absorption spectra as can be seen in Fig. 7.6 by the continuous shift of the bandgap energy with bromine content x.[414]

In order to quantify the apparently increasing exciton binding energy E_B with increasing bromine content x, absorption spectra of $MAPb(I_{1-x}Br_x)_3$ thin films on glass have been recorded for varying iodine–bromine ratio in a temperature range of $T = 5\,K - 277\,K$. Analogously to Fig. 7.1, all spectra have been analyzed by least-squares fits of the generalized Elliott formula. The extracted exciton binding energies E_B are displayed in Fig. 7.7. The accuracy is estimated to be in the range of about 5 meV.

For pure $MAPbI_3$ (dark green squares), the orthorhombic-to-tetragonal phase transition is visible at $T_p \approx 160\,K$ similarly as in Fig. 7.2 (b). In the room-temperature tetragonal phase, E_B is slightly reduced in accordance with previous results.[338,352] All in all, the values for pure $MAPbI_3$ and $MAPbBr_3$ are consistent with published reports and, in particular, coincide with similar spectroscopic studies (see Tab. 7.1).

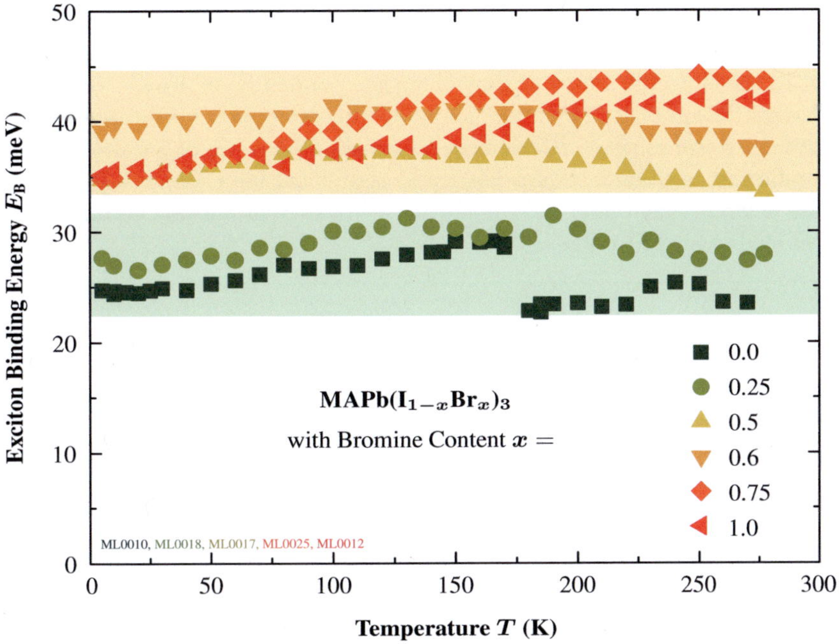

Figure 7.7: Exciton binding energies E_B for $MAPb(I_{1-x}Br_x)_3$ obtained by generalized Elliott fits for $T = 5\,K - 277\,K$. For pure $MAPbI_3$ (dark green squares), the orthorhombic-to-tetragonal phase transition is visible as a jump towards slightly lower values as also seen in Fig. 7.2. Overall, E_B yields larger values for increasing bromine content x in the range of about 25 meV up to 45 meV. Interestingly, there seems to be a small gap in E_B between $x = 0.25$ and $x = 0.5$ (see colored areas).

The obtained exciton binding energies E_B exhibit an overall trend towards larger values for increasing bromine content x (compare to Fig.7.6) starting from slightly above 25 meV (for $x = 0.0$) up to more than 40 meV (for $x > 0.75$). However, the increase appears to be non-linear and (for most temperatures) non-monotonous with bromine content x. In terms of photovoltaic performance, for instance, power-conversion efficiencies exceeding 8% [238,415] and open-circuit voltages up to 1.61 V[416] have been achieved for pure $MAPbBr_3$ ($x = 1.0$) despite the rather high exciton binding energy (e.g., compared to $MAPbI_3$) which possibly hampers charge-carrier separation. This performance results for $MAPbBr_3$-based solar cells are even more notable considering the less suitable bandgap energy ($E_g \approx 2.3\,eV$) for efficient harvesting of the solar spectrum which is reflected by a Shockley-Queisser limit of only ~16%[146] (compared to ~30% at $E_g = 1.6\,eV$ in case of $MAPbI_3$[146]). However, they are still clearly outperformed by $MAPbI_3$-based devices with record efficiencies above 20%.[417]

Interestingly, there seems to be a small gap in E_B between $x = 0.25$ and $x = 0.5$ (see colored areas in Fig. 7.7). Certainly, this needs further detailed investigation taking into account possible uncertainties, e.g., by measuring the exact perovskite composition which has been only estimated from precursor solutions here. Nevertheless, the only slight increase of the exciton binding energy for about 25% and less of incorporated bromine are promising for tandem solar cells since the corresponding bandgap energies E_g of 1.6 eV − 1.8 eV represent optimum values for top cells in combination with silicon or Cu(In,Ga)Se_2-based bottom cells (see chapters 2.1.3 and 2.3)[418–421] Moreover, further engineering of perovskite compositions by incorporation of additional cations has led to significant improvements in terms of device performance and reproducibility.[120, 143, 167, 168, 172, 336] These promising, optimized compounds will be examined in the following section.

7.2.3 The Exciton Binding Energy of Cesium-Containing Multiple-Cation Compounds

Multiple-cation perovskite compounds have not only demonstrated notable stability improvements and enhanced reproducibility compared to methylammonium-only based perovskites, but

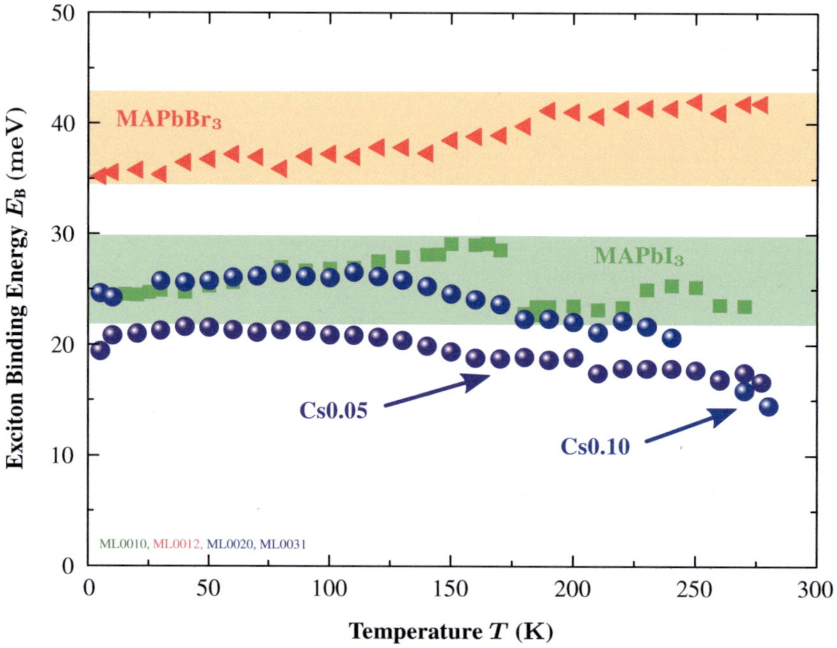

Figure 7.8: Exciton binding energies E_B for two different multiple-cation mixed-halide perovskite compound, namely $Cs_{0.05}(FA_{0.83}MA_{0.17})_{0.95}Pb(I_{0.83}Br_{0.17})_3$ (Cs0.05, purple circles) and $Cs_{0.1}FA_{0.765}MA_{0.135}Pb(I_{0.765}Br_{0.235})_3$ (Cs0.10, blue circles). The results for pure $MAPbI_3$ and $MAPbBr_3$ are shown for comparison (see Fig. 7.7). Both Cs-containing compounds exhibit lower E_B values than $MAPbI_3$, in particular, Cs0.05. From [195], licensed under CC-BY (modified in color).

also represent an important step forward in further increased power-conversion efficiency up to 21.6% and above (see chapter 2).[143, 167, 168, 336] Not only because of the great relevance for solar cells, they are predestined for further material characterization in order to elucidate the effect of different cations and their combinations. Here, especially two different compounds including cesium, formamidinium (FA) and methylammonium (MA) as well as iodine and bromine serve as widely-used photovoltaics-relevant examples: $Cs_{0.05}(FA_{0.83}MA_{0.17})_{0.95}Pb(I_{0.83}Br_{0.17})_3$ (abbreviated as Cs0.05) and $Cs_{0.1}FA_{0.765}MA_{0.135}Pb(I_{0.765}Br_{0.235})_3$ (Cs0.10).

In analogy to the $MAPb(I_{1-x}Br_x)_3$ studies, temperature-dependent absorption spectra have been measured and evaluated using Elliott fits for both Cs-containing mixed perovskites. Figure 7.8 summarizes the results for Cs0.05 (purple circles) and Cs0.10 (blue circles) in comparison with values for pure $MAPbI_3$ (green squares) and $MAPbBr_3$ (red triangles) as discussed before. Both mixed-cation compounds exhibit smaller exciton binding energies compared to $MAPbI_3$ which is especially pronounced in the case of Cs0.05. In addition, the bandgap energies E_g of Cs0.05 and $MAPbI_3$ are very similar around room temperature (1.65 eV and 1.66 eV, respectively) and, therefore, the expected maximum power-conversion efficiency from the Shockley-Queisser limit is nearly identical (opposed to the investigations of $MAPb(I_{1-x}Br_x)_3$ presented above). This indicates that the lower exciton binding energy, which is beneficial for charge-carrier separation and collection, can – at least partly – account for the better solar-cell performance for the multiple-cation compound Cs0.05.

However, it should be noted that, in particular, for small exciton binding energies and large line-width broadening at high temperatures, the excitonic peak in the absorption spectra cannot be reliably separated from the continuum contribution above E_g as can be seen in Fig. 7.9 (The spectra are shifted by the energy position E_{ex} of the excitonic peak maximum and normalized to the absorbance at $E_n = E_{ex} + 0.25\,\text{eV}$.).

	Exciton Binding Energy E_B		
	at 10 K	at 150 K	> 150 K
$MAPbBr_3$	36 meV	38 meV	41 meV
$MAPbI_3$	25 meV	29 meV	32 meV
Cs0.05	21 meV	19 meV	24 meV
Cs0.10	24 meV	25 meV	20 meV

Table 7.2: Summary of determined exciton binding energies E_B for Cs0.05 and Cs0.10 multiple-cation mixed-halide perovskites and pure $MAPbI_3$ and $MAPbBr_3$ for comparison. Values at temperatures of 10 K and 150 K are directly taken from Fig. 7.8 while the last column represents the results of the f-sum rule method for $T > 150\,\text{K}$. The exciton binding energy is reduced for the Cs-containing compounds even compared to $MAPbI_3$ and, in particular, at high temperatures.[195]

Thus, the accuracy of the Elliott fits as well as the extracted E_B values can be compromised (see appendix B). Since this might affect especially the low-E_B Cs-containing perovskites at temperatures approaching room temperature, the so-called f-sum rule method proposed by Sestu *et al.* (introduced in chapter 4.3) has been additionally applied (Here, the integration is performed from $E - E_{ex} = -0.15\,\text{eV}$ to $E - E_{ex} = 0.25\,\text{eV}$.).

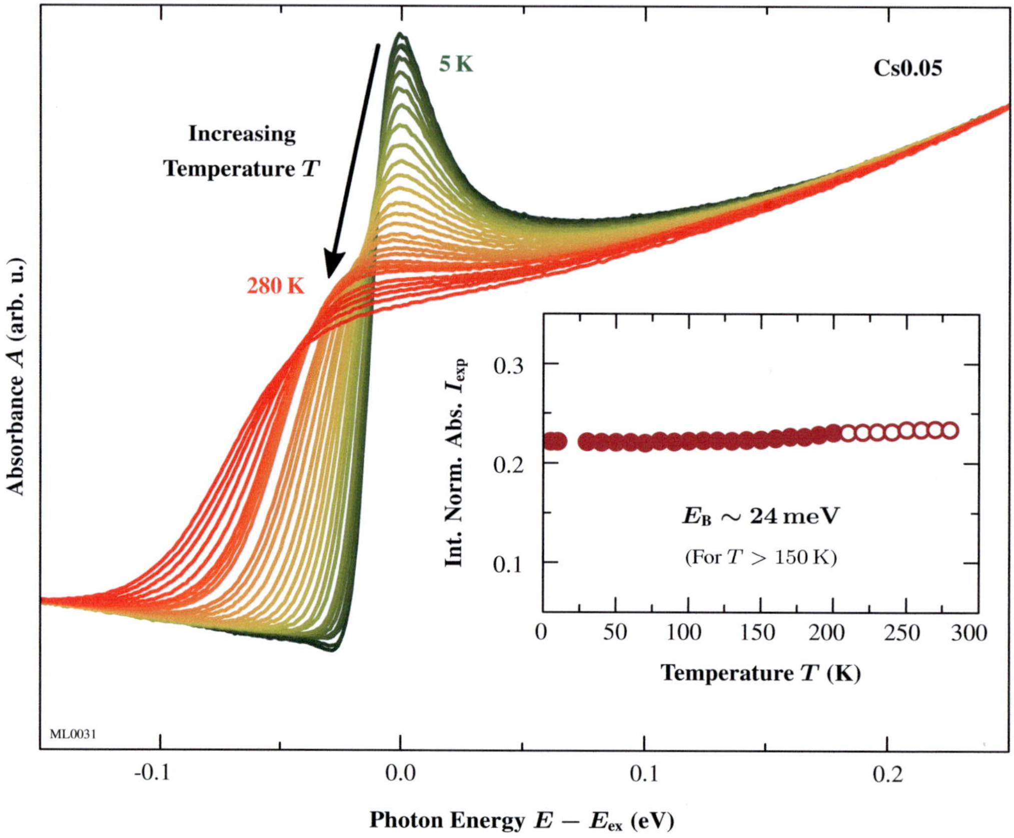

Figure 7.9: f-sum rule method proposed by Sestu *et al.*[352] (see chapter 4.3) applied to the $Cs_{0.05}(FA_{0.83}MA_{0.17})_{0.95}Pb(I_{0.83}Br_{0.17})_3$ (Cs0.05) perovskite compound. The measured absorption spectra are shifted by the energy position E_{ex} of the excitonic peak and normalized to the absorbance at the normalization energy $E_n = 0.25\,\text{eV}$ (For the spectra at $T \geq 210\,\text{K}$, E_{ex} could not be determined accurately and was extrapolated linearly which corresponds to the open circles in the inset.). Inset: The integrated normalized absorption I_{exp}, which has been numerically calculated for the depicted energy range and each temperature T, is nearly constant as expected. I_{exp} corresponds to an exciton binding energy of $E_B \sim 24\,\text{meV}$ for temperatures $T > 150\,\text{K}$ (with $8\mu^2 b/\hbar^4 \approx 1.24\,\text{eV}^{-1}$ estimated from direct Elliott fits). From [195], licensed under CC-BY.

As expected, the resulting broadening-independent integral I_{exp} is mainly constant in the temperature range investigated (see inset of Fig. 7.9). By numerical calculations as explained in chapter 4.3, I_{exp} is mapped to absolute values for the exciton binding energy yielding a value of $E_B \sim 24\,\mathrm{meV}$. Since this technique is especially exploited to improve the reliability of results at temperatures approaching room temperature, the temperature range of the f-sum evaluation is restricted to $T > 150\,\mathrm{K}$. The respective results for Cs0.10, $MAPbI_3$, and $MAPbBr_3$ are summarized in Tab. 7.2. While the values for E_B determined by the f-sum rule method are slightly higher compared to the ones in Fig. 7.8, the overall trend of reduced exciton binding energies for the investigated multiple-cation mixed-halide compositions Cs0.05 and Cs0.10 compared to $MAPbI_3$ and $MAPbBr_3$ is confirmed. The influence of additional bromine and formamidinium in the mixed compounds can be considered to roughly cancel out each other: The incorporation of small amounts of bromine (here 17% and 23%, respectively) leads to an increase of E_B of a few meV as demonstrated in Fig. 7.7. On the other hand, the additional FA slightly reduces E_B as reported, e.g., from magnetoabsorption spectroscopy ($E_B = 10\,\mathrm{meV}$ for $FAPbI_3$ compared to $E_B = 12\,\mathrm{meV}$ for $MAPbI_3$, both at $T \sim 160\,\mathrm{K}$).[407]

With regard to photovoltaic performance, the reduced exciton binding energy of the Cs0.05 and Cs0.10 compounds will favor the separation of electron–hole pairs and, thereby, the collection of electrons and holes in solar cells with the respective perovskite absorber layer. Therewith, it can – at least partly – account for improved solar-cell performance in these nowadays widely-used compounds.

7.3 Implications of Excitonic Effects for Solar-Cell Performance

The clear signature of excitons in optical spectra of organic–inorganic perovskites, especially in electromodulation spectroscopy, raises the question of the consequences for optoelectronic applications, here, with a focus on solar cells. The beneficial effect of lower exciton binding energy on separation of photo-generated electron–hole pairs has already been mentioned in the previous section by comparing different perovskite absorber compositions which can be used for bandgap tuning as well as high-performance solar cells. In the following sections, a more general discussion of excitonic effects and the exciton binding energy on the photovoltaic performance of perovskites will complement this chapter. This includes beneficial effects due to increased absorption, implications for charge-carrier extraction, and the effective expected ratio of excitons and free charge carriers under application-relevant conditions.

The discussion about beneficial absorption effects due to the Sommerfeld enhancement in the following section is partly published in [358]. The temperature-dependent photocurrent measurements in section 7.3.2 were achieved in collaboration with Philip Lange within his master's thesis.[422]

7.3.1 Increased Absorption Coefficient due to Sommerfeld Enhancement

At first glance, excitons seem to be detrimental for high-performance solar cells in any case since bound states of photogenerated electrons and holes are opposed to effective extraction of these charge carriers. While this is true for strongly bound electron–hole pairs, there are indeed advantageous excitonic effects which can predominate for sufficiently small exciton binding energy. In this context, the main point to be mentioned is the increased absorption coefficient in excitonic semiconductors described by the so-called *Sommerfeld enhancement* (see also chapter 3.3.3).[78, 348]

Figure 7.10 visualizes this effect in the case of perovskites (here in $MAPbI_3$). While the solid red lines once more represent least-squares fits of the generalized Elliott formula, the blue and dashed curves are derived from these fits by manually modifying particular parameters. For vanishing excitonic effects described by $E_B = 0$ (dashed lines), the spectra are apparently different. First of all, the discrete excitonic peak below the bandgap energy E_g (dashed black line) vanishes as expected. However, even above the bandgap energy E_g, the spectral shape of the absorption spectrum is apparently different.

In the case of non-vanishing excitonic effects (i.e., nonzero exciton binding energy E_B), even for no longer bound, but ionized excitons above E_g, the residual Coulomb attraction of electrons and holes still leads to a correlation of their statistical positions and, thereby, their wavefunction overlap. This influences the absorption probability compared to negligible excitonic effects and affects the spectral shape of the absorption spectrum. In the ideal case of perfectly parabolic band dispersion (blue curves), the square-root shaped absorption for non-excitonic band-to-band transitions (dashed blue line) is replaced by a nearly constant absorption in the excitonic case (see chapter 3.3)).

More importantly, the magnitude of the absorption coefficient in the ionization continuum is strongly enhanced by these excitonic effects which can be understood as magnified absorption probability or oscillator strength due to increased correlation of electron and hole positions (see equation (3.15) in chapter 3.3). The consideration of the reverse process might offer a more intuitive approach: An increased probability of finding electron and hole at the same position (meaning a higher wavefunction overlap) due to mutual Coulomb interaction results in an enhanced probability of radiative recombination of electrons and holes. Figure 7.10 (b) clearly demonstrates the relevance of this enhancement even at room temperature leading to an absorption increase of more than a factor of 2 for photon energies in the vicinity of E_g. In addition, the non-parabolicity leads to further amplification of the absorption thanks to an increased density of states (see equation (4.2)). These processes contribute to the outstanding absorption coefficient in organic–inorganic perovskites which is one of the key aspects of their favorable properties for photovoltaics enabling the fabrication of thin-film solar cells with an only 300 nm-thin absorber layer. The small layer thickness is especially advantageous for perovskite solar cells in order to facilitate the extraction of charge-carriers within their diffusion length. In that sense, excitonic effects are beneficial for solar-cell performance as long as the exciton binding energy is sufficiently small in order to maintain effective charge-carrier extraction.

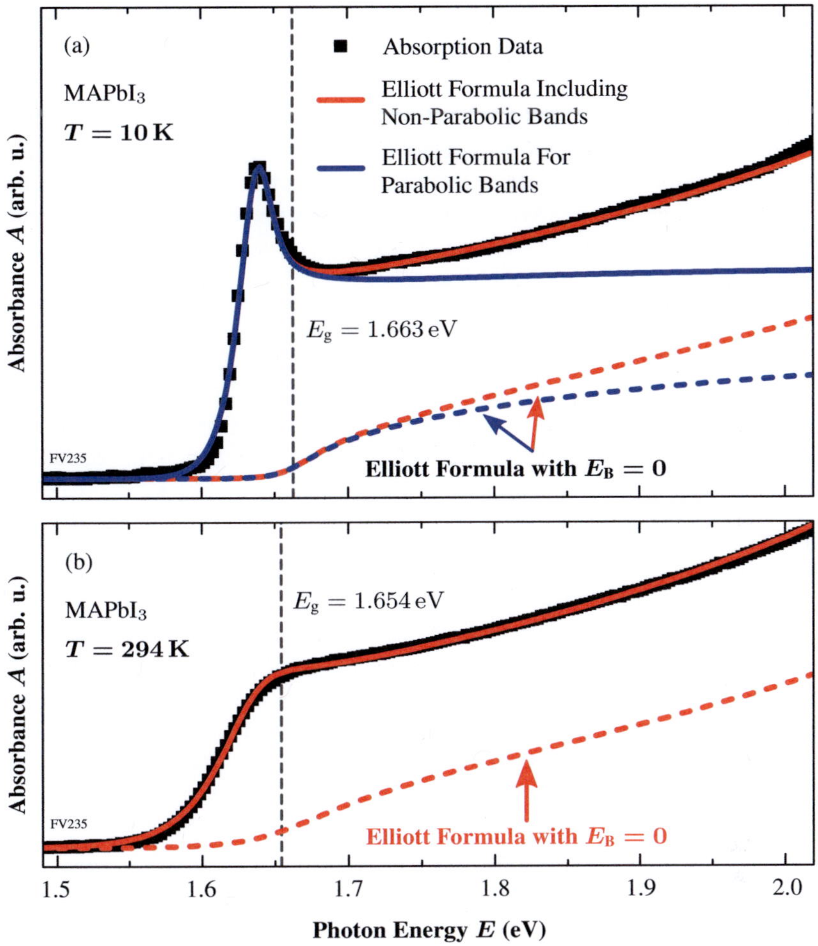

Figure 7.10: Generalized Elliott fits (solid red curves) of measured absorption spectra at (a) $T = 10\,\mathrm{K}$ and (b) room temperature. The blue curves in (a) show the same Elliott fit but with non-parabolicity $8\mu^2 b/\hbar^4 = 0$, i.e., perfectly parabolic bands, to visualize that the increased absorption above E_g is partly related to non-parabolic band effects. The dashed lines represent identical fits but with vanishing exciton binding energy E_B demonstrating the large increase of the absorption coefficient not only at the excitonic peak below E_g, but also in the ionization continuum above E_g. This effect is known as Sommerfeld enhancement and even relevant at room temperature as visible in (b). Original measured spectra obtained in collaboration with Alice Magin.[389] © 2018 IEEE. Reprinted, with permission, from [358].

7.3.2 The Influence of Excitonic Effects on Charge-Carrier Extraction in Perovskite Solar Cells

In order to study the impact of excitons on charge-carrier separation and collection in perovskite solar cells, simultaneous temperature-dependent measurements of absorption and photocurrent spectra have been performed. Details of the measurement principle, experimental setup, and data analysis can be found in appendix A.2 and [422]. For a more straightforward interpretation, the *external quantum efficiency* (EQE) was calculated, i.e., the spectrally resolved ratio of extracted charge carriers per incident photon:[197,211,423,424]

$$\mathrm{EQE}\left(\hbar\omega\right) = \frac{\#\text{ of extracted charge carriers }\left(\hbar\omega\right)}{\#\text{ of incident photons }\left(\hbar\omega\right)} = \frac{n_{\mathrm{e,h}}\left(\hbar\omega\right)}{n_{\mathrm{ph}}\left(\hbar\omega\right)}. \tag{7.1}$$

However, only relative EQE values (in arbitrary units) can be obtained at low temperatures since a quantitative calibration of the incident photon flux per area is not feasible inside the used cryostat.

Fig. 7.11 presents a comparison of absorptance (calculated by $A' = 1 - T$) and (relative) EQE spectrum at temperature $T = 10\,\mathrm{K}$ for $MAPbI_3$. From high photon energies around 1.9 eV to the absorption edge at about 1.6 eV (from right to left), the decline in EQE in (b) is apparently stronger compared to the one in absorptance in (a). Furthermore, the excitonic peak is much less pronounced in external quantum efficiency. This indicates that, as expected, bound electron–hole pairs in the discrete excitonic peak (or with small excess energy close to it) contribute less to the EQE (and, thereby, the photocurrent) than the unbound electrons and holes in the continuum for a comparable number of absorbed photons. Surprisingly, despite the very low thermal energy at $T = 10\,\mathrm{K}$, a significant amount of charge-carriers can be extracted at the excitonic peak. Thus, additional mechanisms have to be responsible for the dissociation of excitons, presumably the dissociation at interfaces similar to organic solar cells.[1,399,425] Patel *et al.* observed similar excitonic effects in temperature-dependent photocurrent spectra but with an even stronger reduction at the excitonic peak for temperatures $T < 45\,\mathrm{K}$. This further demonstrates the impact of excitons on the charge-carrier separation and collection.[426]

Despite the clear signature of reduced contributions of excitons to the photocurrent (and corresponding EQE) at cryogenic temperatures, similar effects are not observable at high temperatures. Moreover, the spectrally integrated (or total) photocurrent at room temperature is more than three orders of magnitude larger compared to the one at $T = 10\,\mathrm{K}$ as shown in Fig. A.3 in chapter A.2 (similar to other reports[426,427]) rationalizing the experimentally obtained high power-conversion efficiencies of perovskite solar cells under normal operation conditions.

In accordance with the room-temperature observations, there is a general consensus in literature that charge-carrier dynamics in perovskites are dominated by free charge carriers at solar-cell conditions, i.e., room temperature and 1 sun illumination.[20] Accordingly, the fast dissociation of excitons into free electrons and holes is reflected by experimentally obtained short dissociation times in the order of picoseconds.[428,429]

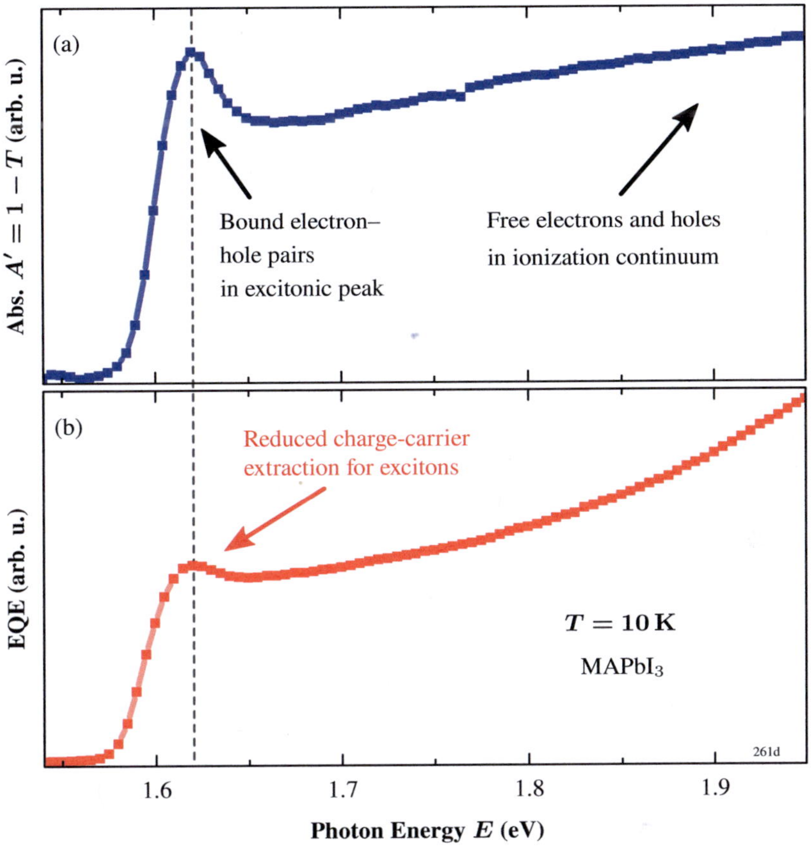

Figure 7.11: Simultaneous measurements of (a) absorptance $A' = 1 - T$ and (b) relative external quantum efficiency (calculated from photocurrent spectra) of $MAPbI_3$ at $T = 10\,\mathrm{K}$. As expected, the contribution to the EQE (and, thereby, the photocurrent) is reduced in the discrete excitonic peak of bound electron–hole pairs at $E = 1.62\,\mathrm{eV}$ (dashed line) compared to the ionization continuum at higher photon energies. Data published in [422].

The Saha–Langmuir equation, originally derived for the ionization of atoms in stars,[430–432] provides a simple illustration of this by considering a gas of excitons and free charge carriers in equilibrium (using a mass-action law).[79, 412, 433–435] The ratio $x = n_{\mathrm{free}}/n$ of free charge-carriers n_{free} (compared to excitons n_{exc}) to the total excitation density $n = n_{\mathrm{free}} + n_{\mathrm{exc}}$ is then given by[412, 435]

$$\frac{x^2}{1-x} = \frac{1}{n}\left(\frac{2\pi\mu k_{\mathrm{B}}T}{h^2}\right)^{3/2} e^{-\frac{E_{\mathrm{B}}}{k_{\mathrm{B}}T}} \tag{7.2}$$

with reduced exciton mass μ (approximated to be $0.15\,m_{\mathrm{e}}$[340, 412]), Boltzmann and Planck constant k_{B} and h, respectively, and temperature T.

Figure 7.12 illustrates the results for different exciton binding energies E_B at room temperature in (a), and varying temperature for a fixed intermediate $E_B = 30\,\mathrm{meV}$ in (b). The photovoltaic regime can be estimated by the total absorbed solar flux and the excitation lifetime to be $10^{13}\,\mathrm{cm}^{-3} - 10^{15}\,\mathrm{cm}^{-3}$ and it is strongly dominated by free charge carriers at room temperature ($x > 90\%$) even for the highest reported exciton binding energies (see Fig. 7.12 (a)).[412] As expected, the fraction of excitons is strongly increased at lower temperatures. Nevertheless, under different conditions, both excitons and free charge carriers – or even both simultaneously – have been observed at room temperature using various measurement techniques, e.g., electromodulation spectroscopy as presented in section 7.1.[76,415,436]

Certainly, the presented model is rather simplified and should only be seen as an illustrative description. For instance, exciton–exciton interaction and the formation of an electron–hole plasma at high excitation densities are not included (see, e.g., [344, 434]).

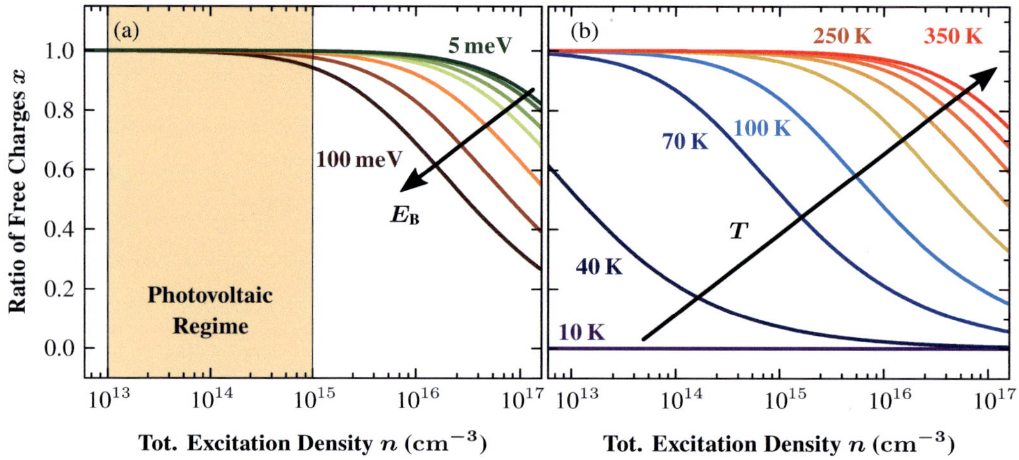

Figure 7.12: Illustration of the fraction $x = n_{\mathrm{free}}/n$ of free charge carriers n_{free} for different total excitation densities $n = n_{\mathrm{free}} + n_{\mathrm{exc}}$. (a) Varying exciton binding energy E_B at room temperature. The photovoltaic regime of $n \approx 10^{13}\,\mathrm{cm}^{-3} - 10^{15}\,\mathrm{cm}^{-3}$ is strongly dominated by free charge carriers even at the highest E_B values reported in literature. (b) Varying temperature T for a fixed intermediate $E_B = 30\,\mathrm{meV}$. At low temperatures, excitonic effects become significant. Visualization based on [412].

7.4 Summary

In this chapter, the role of excitons in organic–inorganic perovskite solar cells for both, fundamental optical transitions and photovoltaic performance of different perovskite compounds, is examined in detail. The combined evaluation of simultaneously recorded, temperature-dependent electroabsorption and standard absorption spectra proves the excitonic nature of fundamental transitions in the mostly-used perovskite model system $MAPbI_3$. The resonance energy in $\Delta T/T$ electroabsorption spectra is clearly below the bandgap energy (determined by generalized Elliott fits) and agrees well with the energy position of the excitonic peak in the corresponding absorption spectra. This demonstrates that the fundamental transitions probed by electromodulation spectroscopy arise from discrete states of excitons opposed to unbound electron–hole pairs in the ionization continuum above the bandgap energy. This is found to be true for the whole temperature range investigated from $T = 10\,\mathrm{K}$ up to room temperature. Consequently, resonance energies extracted from electromodulation spectra do not coincide with the bandgap energy as compared to other solar-cell materials.[74,75] Moreover, the best agreement of experimental $\Delta T/T$ spectra with a Gaussian-broadened FDFF line shape expected for excitonic transitions further substantiates these conclusions.

In order to compare different perovskite compositions, the exciton binding energy E_B is determined by least-squares fits of the generalized Elliott formula[77,79] to experimental temperature-dependent absorption spectra and the f-sum rule method.[352] For varying iodine–bromine ratio x in $MAPb(I_{1-x}Br_x)_3$, which can be utilized for bandgap tuning, a (non-linear) increase of E_B from about 25 meV up to 45 meV with increasing bromine content is observed. Additional incorporation of cesium and formamidinium on the cation site in compounds with low bromine content (17% or 23.5%) leads to a further reduction of the exciton binding energy even below the values for pure $MAPbI_3$. These low exciton binding energies of about 20 meV in multiple-cation compounds are favorable for charge-carrier separation and collection in solar cells.

Furthermore, implications of excitonic effects on the performance of perovskite solar cells in general are discussed: The exciton binding energy E_B in the order of thermal energies at room temperature leads to a significantly increased absorption coefficient compared to vanishing E_B (more than a factor of 2 for photon energies close to E_B) even at room temperature. This so-called Sommerfeld enhancement is a consequence of residual Coulomb attraction between unbound electrons and holes in the ionization continuum above the bandgap energy. With respect to the influence on charge-carrier extraction, temperature-dependent photocurrent spectroscopy indeed reveals that, as expected, at low temperatures, electron–hole pairs bound as excitons contribute less to the extracted photocurrent compared to unbound electrons and holes in the ionization continuum. However, the difference is surprisingly small and, in particular, only observable at temperatures far below room temperature. This is also illustrated by the so-called Saha–Langmuir equation which predicts ratios of free charge-carriers compared to excitons above 90% in perovskites for illumination intensities in the photovoltaic regime at room temperature.[412,430,432]

In conclusion, excitonic effects play a key role in organic–inorganic perovskites which combine the desirable properties for photovoltaics, being high absorption coefficient and efficient charge-carrier extraction. On the one hand, the notable exciton binding energy leads to a significant increase in absorption due to the Sommerfeld enhancement. But, on the other hand, it is still small enough to ensure efficient dissociation of excitons and extraction of electrons and holes. As a result, the remarkably high power-conversion efficiencies of perovskite solar cells exceeding 20% can be rationalized, which have been reported e.g., for Cs-containing multiple-cation compounds. These promising candidates for high-performance devices possess especially favorable characteristics and will be investigated in more detail with regard to temperature-induced crystal-phase transition (chapter 8) and compositional stability (chapter 9).

Chapter 8

Crystal-Phase Transitions in Organic–Inorganic Perovskites

Organic–inorganic perovskites such as the model system $MAPbI_3$ do not only exist in the often depicted cubic crystal structure, but possess a lower-symmetry tetragonal or orthorhombic structure at reduced temperatures. Transitions between these different crystal phases are caused by thermally-induced rotation and orientation of the organic cation (e.g., MA) leading to distortions of the surrounding inorganic (Pb–I) framework (see chapter 2.1.1).

Since the occurrence of crystal phase transitions could influence the performance of perovskite-based devices, e.g., solar cells or lasers, in-depth studies are not only interesting from a materials-characterization point of view, but are also relevant in terms of applications. For instance, Jia *et al.* attribute their achievement of continuous-wave lasing in $MAPbI_3$ to charge-carrier accumulation caused by a mixed orthorhombic-tetragonal phase at low temperatures.[39] However, in general, suppression of temperature-induced phase transitions is preferred for stable performance of devices,[42] e.g., in photovoltaic applications in space where temperatures between 90 K and 350 K are not uncommon.[437]

Temperature-induced (and phase-transition related) changes of the average positions of the atoms relative to each other modify their wave function overlap and, thereby, the electronic structure of the material, e.g., the bandgap energy E_g. This opens up the possibility of spectroscopic detection of crystal-phase transitions. In this chapter, temperature-dependent absorption and electromodulation spectroscopy are utilized to study phase transitions in $MAPbI_3$ and $MAPbBr_3$ (section 8.1) as well as in $Cs_xFA_yMA_{1-x-y}Pb(I_{1-z}Br_z)_3$ compounds (section 8.2) introduced in the previous chapter (Cs0.05 and Cs0.10). In addition, the achieved spectroscopic results are verified by temperature-dependent *X-ray diffraction* (XRD) results (section 8.3).

8.1 Phase Transitions in Methylammonium Lead Iodide and Methylammonium Lead Bromide

The results in the following sections are published in [195, 358]. Electromodulation spectra and resulting resonance energies have been obtained in collaboration with Alice Magin within her master's thesis.[389]

In case of methylammonium lead iodide, $MAPbI_3$, the transition between the orthorhomic crystal phase below $T_p \approx 160\,K$ and the room-temperature tetragonal phase is clearly detectable in temperature-dependent optical spectra. $\Delta T/T$ electroabsorption spectra depicted in Fig. 8.1 are utilized for an exemplary discussion of the key features.

In addition to the expected increase of spectral broadening, the resonance features in the $\Delta T/T$ spectra overall shift towards higher photon energies with increasing temperature which is a result of the temperature dependence of the perovskite's bandgap energy with $dE_g/dT > 0$. This behavior is in striking contrast to most typical semiconductors such as GaAs or Ge, which exhibit a decrease of the bandgap energy with increasing temperature.[438–442] The reason for this is found in the specific band structure of lead-based perovskites and, in particular, the antibonding nature of the valence band (see chapter 2.1.2).[81, 412, 443] The thermal expansion of the lattice with temperature decreases the overlap of the Pb–6s and I–5p orbitals forming the valence band. Consequently, the valence band maximum (VBM) shifts downwards in energy.[443, 444] Since this shift exceeds the changes of the conduction band minimum (CBM), the bandgap energy increases with temperature which has been shown theoretically and experimentally.[408, 443–445] For instance, increasing the temperature from 25 °C to 85 °C leads to a decrease of 110 meV and 77 meV of VBM and CBM, respectively, in experiments resulting in an increase of the bandgap energy of 33 meV.[444]

Furthermore, the pronounced abrupt shift of the spectra of about 100 meV towards lower energies around $T = 160\,K$ marks the transition from the low-temperature orthorhombic to the room-temperature tetragonal crystal phase.[412, 443, 446] For the understanding of temperature-induced changes in the different crystal phases, the orientation of the organic A-site cation, here $CH_3NH_3^+$ (MA), plays a prominent role (see chapter 2.1.1). The orthorhombic crystal structure at low temperatures is characterized by the fully (antiparallel) aligned MA cations due to hydrogen bonding between the MA and the inorganic Pb–I cage.[82, 84, 85] As a result the inorganic Pb–I framework is distorted and the PbI_6 octahedra are tilted (see chapter 2.1.1).[82, 84–86] This distortion of the Pb–I bonds is reduced in the tetragonal phase above $T_p \approx 162\,K$[80] due to increased thermal motion of the MA cation and, thereby, less hydrogen bonding.[82, 84] Theoretical calculations show that the reduced distortion of Pb–I bond angles leads to a decrease of the bandgap energy which is clearly validated by optical spectroscopy.[81, 96, 102, 447]

Moreover, another important feature accompanying the orthorhombic-to-tetragonal phase transition is the coexistence of both phases in a certain temperature range which show up as two concurrent resonance features in the $\Delta T/T$ spectrum at 150 K in Fig. 8.1.[84, 443, 446, 448] This is a result of the spatially non-uniform transition process leading to tetragonal-phase inclusions in the orthorhombic phase or vice versa over a certain temperature range and is probably related to the grain structure of the perovskite layers.[89, 331, 443, 446, 448, 449] As a consequence, the observed macroscopic phase-transition temperature depends, in particular, on the temperature profile (measured from low to high temperatures or vice versa). This effect is known as "temperature hysteresis".[449] In extreme cases, i.e., under local laser excitation, tetragonal phase inclusions have been observed even at 5 K.[448]

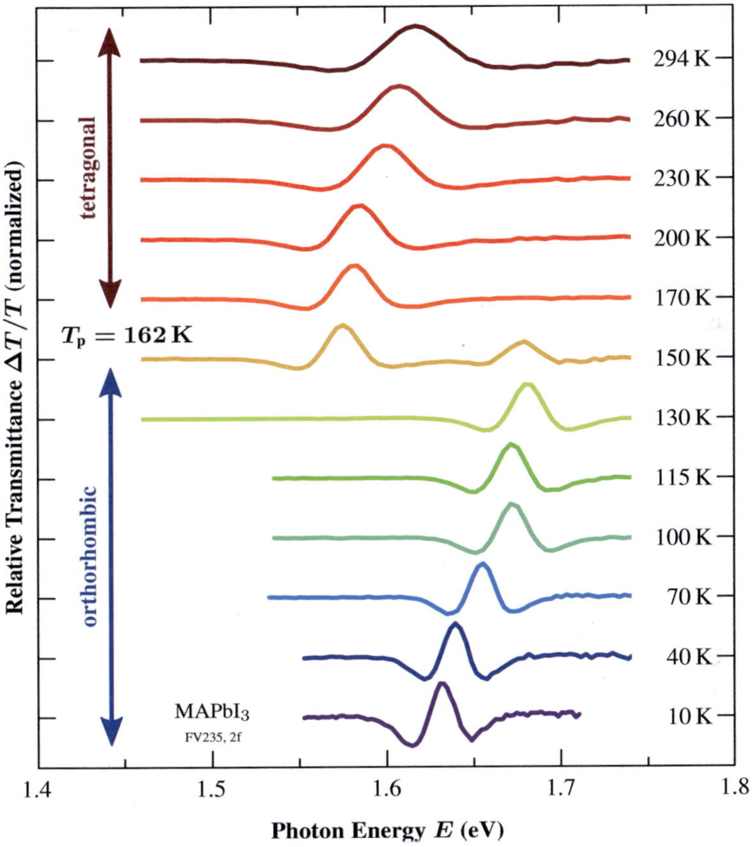

Figure 8.1: Temperature-dependent electroabsorption spectra of a semitransparent solar cell with $MAPbI_3$ absorber layer for $T = 10\,K - 294\,K$.[358] The shift of the spectra towards higher energies with increasing temperature is a result of the temperature dependence of the bandgap energy with $dE_g/dT > 0$ which is typical for lead-based perovskites due to their antibonding valence bands.[81, 412] The orthorhombic-to-tetragonal phase transition reveals itself by an abrupt shift towards lower photon energies around the phase-transition temperature $T_p = 162\,K$.[80] The coexistence of both phases at 150 K is related to so-called "temperature hysteresis" effects and caused by small inclusions of both phases on the microscale. © 2018 IEEE. Reprinted, with permission, from [358].

Another straightforward way for spectroscopic studies of crystal phase transitions without the need of fabricating complete devices is the measurement of temperature-dependent absorption spectra of perovskite thin films on glass substrates. The subsequent evaluation can be performed by generalized Elliott fits similar to the determination of the exciton binding energy E_B in chapter 7.2. In fact, the results presented in the following complement the analysis in that chapter. Since E_B has been shown to depend only weakly on temperature, extracted bandgap energies E_g as depicted in Fig. 8.2 provide reliable results.

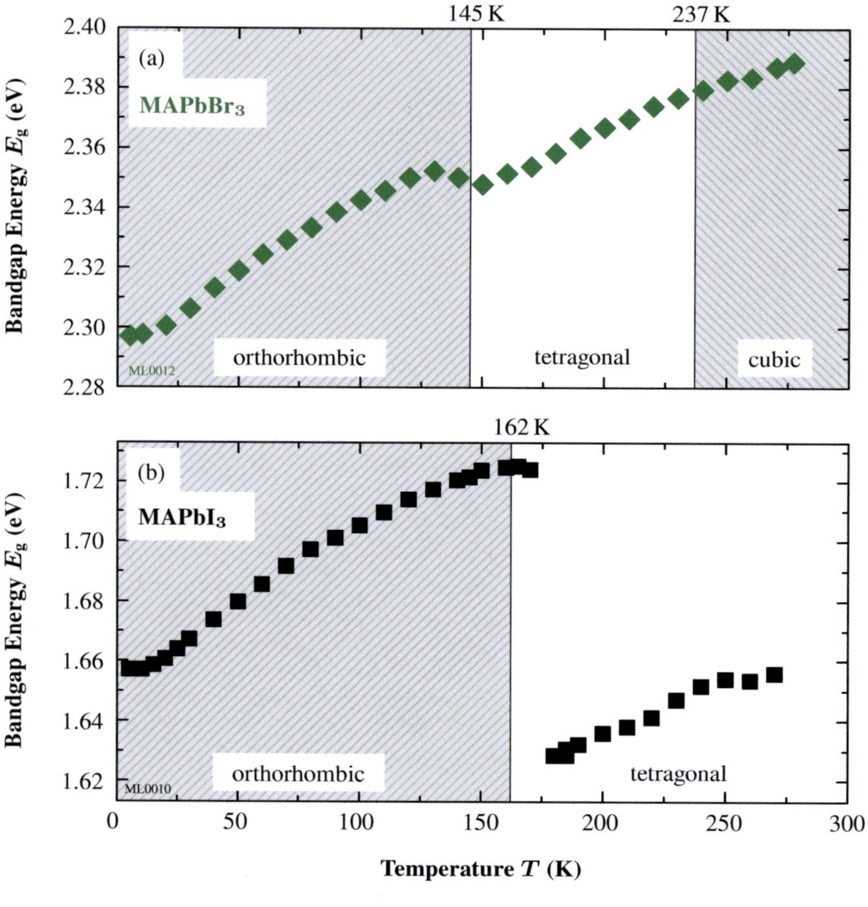

Figure 8.2: Temperature dependence of the bandgap energy E_g of (a) $MAPbBr_3$ (green diamonds) and (b) $MAPbI_3$ (black squares) extracted from generalized Elliott fits (related to Figs. 7.6 and 7.7). The orthorhombic-to-tetragonal phase transition is clearly marked by a kink or jump ($\sim 100\,\mathrm{meV}$) of E_g around the transition temperature of 145 K[80] and 162 K,[80] respectively. However, no clear indication is observed for the tetragonal-to-cubic transition at 237 K.[80] From [195], licensed under CC-BY (selected data).

In case of $MAPbI_3$, the abrupt shift of $\sim 100\,\mathrm{meV}$ at temperature $T \approx 160\,\mathrm{K}$ in Fig. 8.1 indicating the orthorhombic-to-tetragonal phase transition can be rediscovered in the temperature dependence of E_g extracted from absorption spectra which is shown in Fig. 8.2 (b). The direct comparison of absorption and electroabsorption results is discussed in chapter 7.1.1.

The situation is different for $MAPbBr_3$ depicted in Fig. 8.2 (a): In this case, no distinct jump appears at the phase-transition temperature ($T_p \approx 145\,\mathrm{K}$ for $MAPbBr_3$), but rather a small kink-like decrease of E_g.[80, 443, 450, 451] Interestingly, Dar *et al.* observed a similar feature in their photoluminescence spectra for $FAPbBr_3$ but not for $MAPbBr_3$.[443] Clearly, further investigations are required to reveal the underlying effects and the influence of different compositions.

It should be noted that the tetragonal-to-cubic phase transition occurs around 237 K for $MAPbBr_3$[80] which is within the investigated temperature range up to room temperature. However, in this case, no clear indication can be identified in the temperature dependence of the bandgap energy. This observation is in good agreement with previous reports about the influence of the tetragonal-to-cubic phase transition on optical and electronic properties in both $MAPbI_3$ and $MAPbBr_3$. So far, only small changes in photocurrent for $MAPbI_3$[452] and Hall mobility in case of $MAPbBr_3$[453] have been observed, but no influence on the optical bandgap could be detected.[444, 445, 454] Overall, the tetragonal-to-cubic phase transition is found to be much more "continuous" in most parameters and, in particular, lattice constants compared to the orthorhombic-to-tetragonal one.[82, 84, 454–456]

Concerning the implications on photovoltaic performance, no or only minor effects have been found in different studies for the tetragonal-to-cubic transition opposed to clear detrimental effects of the transition from the tetragonal to the orthorhombic phase. These are mainly related to a reduced photocurrent[426, 427, 452] and can become relevant, e.g., in space applications.[437]

8.2 Phase-Transition Suppression in Cesium-Containing Mixed-Cation Compounds

Building on this knowledge about possible changes of bandgap energy and resonance features in absorption and electromodulation spectroscopy, respectively, indicating phase transitions in perovskites, the focus in the following sections is on the Cs-containing multiple-cation compounds Cs0.05[1] and Cs0.10[2] introduced in chapter 7.2.3. The results for the temperature dependence of the bandgap energy E_g extracted from absorption spectra is shown in Fig. 8.3 for both multiple-cation compounds as well as $MAPbI_3$ for comparison (see Fig. 8.2). Concerning the variations for Cs0.05 close to room temperature, the possibly reduced reliability of the Elliott fits has already been discussed in chapter 7.2. Clearly, the bandgap energy of Cs0.10 is overall increased by about 70 meV which is mainly caused by the higher content of cesium and bromide.[157, 457]

[1] $Cs_{0.05}(FA_{0.83}MA_{0.17})_{0.95}Pb(I_{0.83}Br_{0.17})_3$
[2] $Cs_{0.1}FA_{0.765}MA_{0.135}Pb(I_{0.765}Br_{0.235})_3$

More importantly, both Cs-containing compounds do not exhibit any indication of a phase-transition related jump or kink in bandgap energy E_g over the whole temperature range investigated up to room temperature. Consequently, the spectroscopic data suggests a suppression of phase transitions opposed to the clearly observable orthorhombic-to-tetragonal one in $MAPbI_3$ and $MAPbBr_3$ for these temperatures.

In order to further substantiate these conclusions from absorption spectroscopy, temperature-dependent electroreflectance spectra of complete solar cells with Cs0.05 and $MAPbI_3$ absorber layer were recorded. Thereby, the influence of phase-transitions on the fundamental optical resonance of the perovskite can be probed in a standard layer stack for photovoltaics consisting of FTO/TiO_2/perovskite/spiro-OMeTAD/Au.

The results from electroreflectance spectroscopy for Cs0.05, which is more widely-used in photovoltaics than Cs0.10, compared to $MAPbI_3$ are presented in Fig. 8.4. Similarly to the analysis of absorption spectra in Fig. 8.3, $MAPbI_3$ exhibits a pronounced jump of the resonance energy of about 100 meV around $T \approx 160\,\mathrm{K}$ which is not present in the Cs0.05 compound. Therefore, the conclusion of phase-transition suppression in Cs-containing multiple-cation perovskites is confirmed not only for thin films on glass but also in complete solar cells.

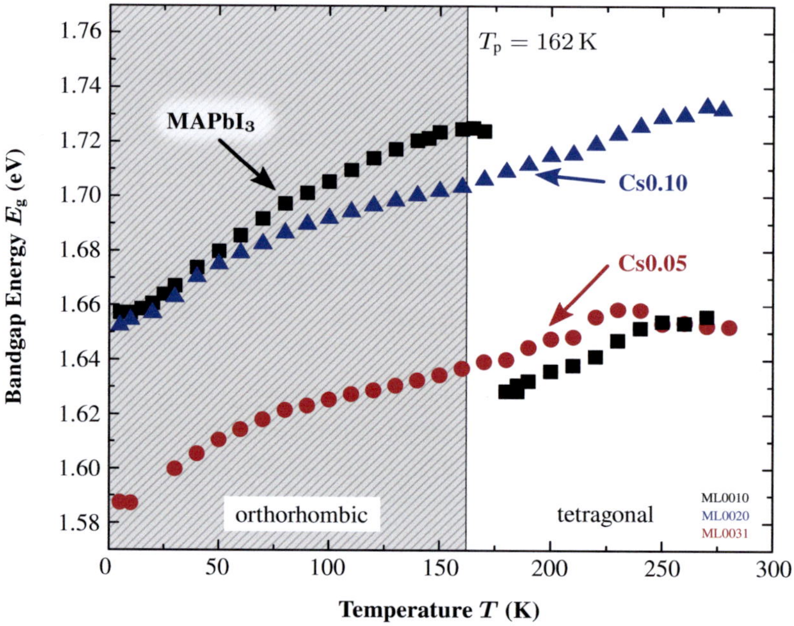

Figure 8.3: Temperature dependence of the bandgap energy E_g for Cs0.05[1] (red circles) and Cs0.10[2] (blue triangles) compounds in comparison with $MAPbI_3$ (black squares, taken from Fig. 8.2) extracted from generalized Elliott fits (related to Fig. 7.8). The absence of any jump or kink for both Cs-containing compounds indicates the suppression of phase transitions in the investigated temperature range. From [195], licensed under CC-BY.

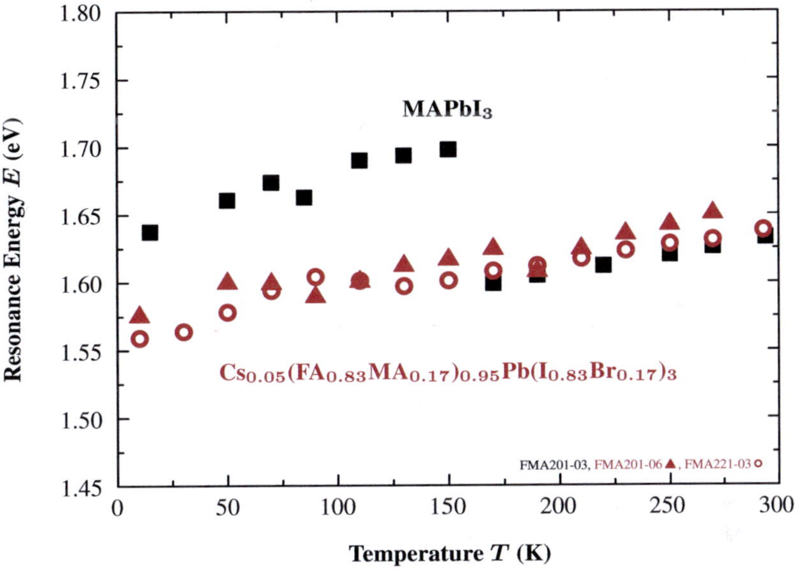

Figure 8.4: Temperature dependence of the fundamental resonance energy of Cs0.05 (red symbols, triangles and circles denote different samples) and $MAPbI_3$ (black squares) obtained by electroreflectance spectroscopy. Similarly to the bandgap energy E_g extracted from absorption spectra in Fig. 8.3, the Cs0.05 compound does not exhibit any jump or kink of the resonance energy (as compared to $MAPbI_3$) confirming the clear indication of phase-transition suppression for $T = 10\,K - 295\,K$ also in complete solar cells. From [195], licensed under CC-BY (selected area).

8.3 Structural Verification of Phase-Transitions by Temperature-Dependent X-Ray Diffraction

Temperature-dependent optical spectroscopy is a very feasible tool for studies of phase-transition induced changes of material properties such as the bandgap energy E_g in both bare absorber layers and complete devices. The application of these techniques on perovskites enabled the identification of the orthorhombic-to-tetragonal phase transition in $MAPbI_3$ and $MAPbBr_3$ as well as its suppression in multiple-cation compounds. However, only structural characterization methods such as *neutron* or *X-ray diffraction* provide unambiguous evidence for crystal structure changes due to phase transitions since alterations in the crystal phase do not translate in clearly observable optoelectronic changes in any case. For instance, the tetragonal-to-cubic transition was not detectable in absorption data of $MAPbBr_3$ in Fig. 8.2.

Consequently, the conclusions drawn from optical spectroscopy are verified by temperature-dependent *X-ray powder diffraction* (XRD) measurements to complement the analysis in this chapter. The XRD patterns have been measured by Bettina Rendenbach in collaboration with Meltem F. Aygüler and Nadja Giesbrecht at Ludwig-Maximilian-University (LMU) in Munich and are published in [195]. The samples are fabricated as thin films on glass, scraped off afterwards, and prepared with grease between two foils. For comparison, XRD diffractograms of grease only are shown in Fig. 8.6 (a) and can account for, e.g., the peak at 13°. Further Details on the experimental system can be found in Appendix E.1.

Figure 8.5 presents the diffractograms for $MAPbI_3$ and $MAPbBr_3$ for temperatures $T = 20\,\mathrm{K} - 300\,\mathrm{K}$. The orthorhombic-to-tetragonal phase transition in $MAPbI_3$ between 140 K and 180 K is clearly apparent by changes in the XRD patterns in Fig. 8.5 (a). For instance, for temperatures $T = 180\,\mathrm{K}$ and above, two additional peaks (marked in gray) appear around 24° which can be assigned to the tetragonal crystal structure (as (211) and (202) reflexes[458]) by comparison with theoretically calculated patterns. Similiarly, specific peaks around 32° (highlighted in gray) are only present either below $T = 180\,\mathrm{K}$ or at $T \geq 180\,\mathrm{K}$ and, thereby, indicate a crystal-phase transition in this temperature range.

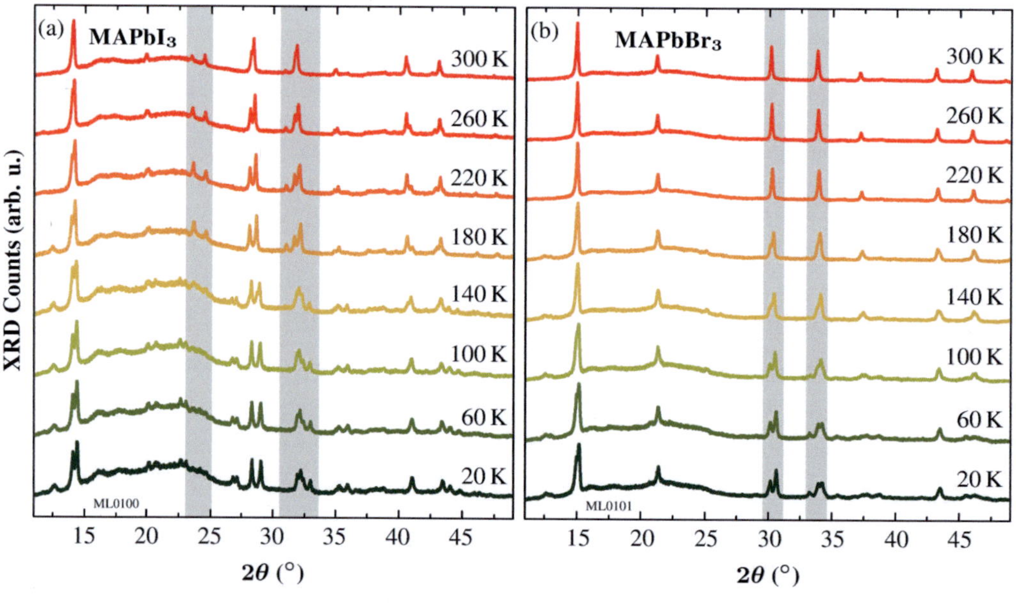

Figure 8.5: Temperature-dependent X-ray powder diffractograms of (a) $MAPbI_3$ and (b) $MAPbBr_3$ for $T = 20\,\mathrm{K} - 300\,\mathrm{K}$ (green to red curves). For both perovskite compositions, temperature-induced changes in the XRD patterns due to crystal-phase transitions are clearly visible (see, e.g, peaks highlighted in gray). In particular in $MAPbI_3$, the orthorhombic-to-tetragonal transition is very well pronounced as observable, e.g., by comparison of diffractograms at $T = 140\,\mathrm{K}$ and $T = 180\,\mathrm{K}$. From [195], licensed under CC-BY (peaks highlighted).

Overall, the diffractograms are in good agreement with theoretical predictions and observed reflexes can be assigned to the low-temperature orthorhombic phase or room-temperature tetragonal phase.[458]

In case of $MAPbBr_3$ shown in Fig. 8.5 (b), the respective temperature-induced variations of the XRD patterns are less pronounced and appear to be more gradual in a range of $T = 100\,K - 220\,K$ (see peaks highlighted in gray) pointing towards a stronger temperature hysteresis effect (see section 8.1). Still, the observed changes reflect the orthorhombic-to-tetragonal phase transition. Interestingly, the tetragonal-to-cubic transition between 220 K and 260 K cannot be detected by changes in the diffractograms and better techniques would be needed for clarification.[80, 455, 459]

Compared to the distinctive phase-transition related changes in X-ray diffractograms of $MAPbI_3$ and $MAPbBr_3$, respective temperature-dependent XRD patterns for Cs0.05 do not exhibit any equivalent characteristics as shown in Fig. 8.6 (a). Only, thermal expansion of the lattice with

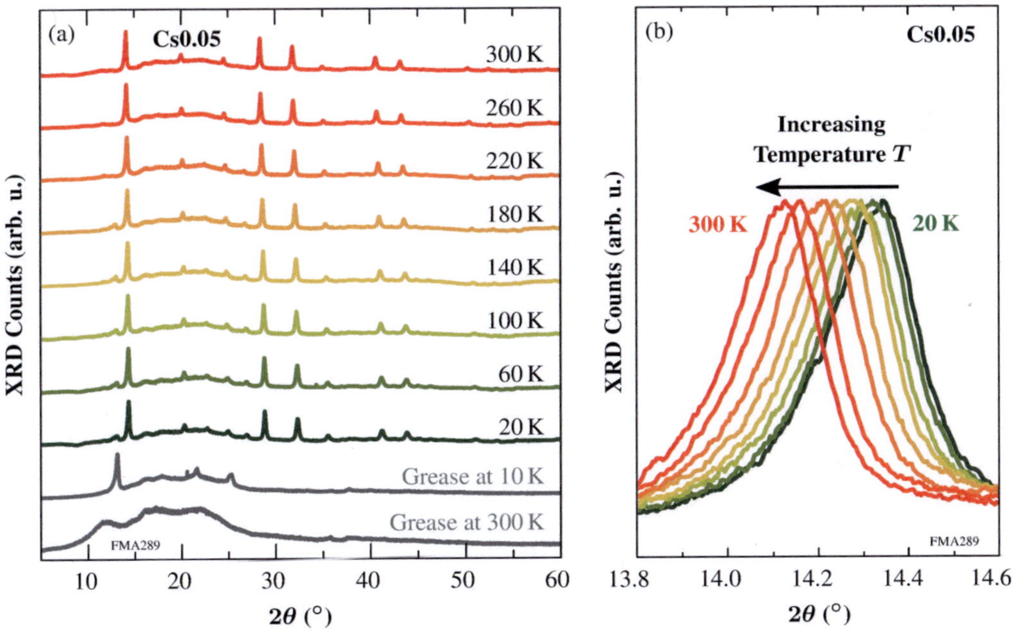

Figure 8.6: (a) Temperature-dependent X-ray powder diffraction patterns of Cs0.05 for $T = 20\,K - 300\,K$ (green to red curves). Diffractograms of grease (grey curves) used for sample preparation at 10 K and 300 K, respectively, are depicted for comparison. The absence of any changes in the perovskite diffractograms – except for a continuous shift caused by temperature-induced lattice expansion (see (b)) – confirms the phase-transition suppression observed in optical spectroscopy. (b) Enlargement of the perovskite peak at $2\theta \sim 14.2°$. As expected, lattice expansion with increasing temperature leads to a continuous shift of the diffraction peak towards lower angles. From [195], licensed under CC-BY.

increasing temperature leads to a shift of the XRD peaks toward lower diffraction angles as expected (see enlargement in Fig. 8.6 (b)). The absence of any other temperature-induced changes perfectly validates the conclusion of phase-transition suppression in Cs0.05 for temperatures up to room temperature. Thereby, not only the feasibility of optical spectroscopy for phase-transition studies is demonstrated but also the applicability of Cs-containing multiple-cation compounds for phase-stable operation in devices such as perovskite lasers or photovoltaic applications in space.[35, 40, 42, 298–300, 460]

Unfortunately, the complexity of these compounds including both multiple cations and multiple halides strongly complicate the theoretical understanding of the underlying processes. Accordingly, no theoretical calculations are reported in literature yet which is probably because of the large computational effort needed for mixed compounds. Therefore, only comparisons with pure compounds are possible, e.g., $CsPbI_3$ which exhibits similarities in structural features and bandgap energy compared to $MAPbI_3$ (but with an orthorhombic crystal structure at room temperature).[139, 461–463]

8.4 Summary

In this chapter, crystal phase transitions in the perovskite compounds $MAPbI_3$, $MAPbBr_3$ and $Cs_xFA_yMA_{1-x-y}Pb(I_{1-z}Br_z)_3$ (in particular, Cs0.05 and Cs0.10) have been studied in detail using absorption and electromodulation spectroscopy. The transition from the low-temperature orthorhombic to the room-temperature tetragonal phase in $MAPbI_3$ can be unambiguously identified as an abrupt decrease of the bandgap energy (extracted from generalized Elliott fits to absorption spectra) or optical resonance energy (obtained from electromodulation spectra). Similarly, a kink of the temperature-dependent bandgap energy is observed at the orthorhombic-to-tetragonal phase-transition temperature $T_\mathrm{p} = 145\,\mathrm{K}$ for $MAPbBr_3$.

Compared to $MAPbI_3$ and $MAPbBr_3$, the Cs-containing perovskite compounds do not reveal any phase-transition related features in bandgap or resonance energy in the investigated temperature range from $T = 10\,\mathrm{K}$ up to room temperature. This clearly indicates the phase-transition suppression in these compounds.

Both the orthorhomic-to-tetragonal phase transition in $MAPbI_3$ and $MAPbBr_3$ and its suppression in Cs0.05 are verified by temperature-dependent X-ray powder diffraction. This emphasizes the applicability of optical spectroscopy for facile investigation of phase transitions in perovskites and proves the suitability of Cs-containing multiple-cation compounds for phase-stable operation in applications such as lasers or in space photovoltaics.[35, 40, 42, 298–300, 460]

Chapter 9

Bandgap Instabilities in Multiple-Cation Mixed-Halide Perovskite Solar Cells

Stability is undoubtedly one of the key challenges of perovskite-based solar cells (and other applications) including both the perovskite material itself and additional layers in the device architecture. As demonstrated in previous chapters, perovskite compounds consisting of mixtures of different ions offer great possibilities of tuning desired properties. Unfortunately, they are predestined for instabilities such as demixing. The most prominent example in this context is the segregation into iodine-rich and bromine-rich phases in $MAPb(I_{1-x}Br_x)_3$ which occurs in seconds under illumination (see section 9.1).

Perovskites additionally incorporating multiple cations do not only possess beneficial properties regarding exciton binding energy and phase-transition suppression as presented in chapters 7 and 8, but are also reported to be stable against halide segregation.[143, 167, 336, 337] However, by utilizing electroreflectance spectroscopy, reversible instabilities of the bandgap energy are found in $Cs_{0.05}(FA_{0.83}MA_{0.17})_{0.95}Pb(I_{0.83}Br_{0.17})_3$[1] under illumination and applied bias simulating solar-cell operation conditions (section 9.2). Investigations of the role of different incorporated ions as well as structural characterization using *in-situ X-ray diffraction* and *Time-of-Flight Secondary Ion Mass Spectrometry* (ToF-SIMS) provide insights into the origin of these instabilities (section 9.3). Subsequently, the influence of illumination and bias as driving forces are thoroughly studied (section 9.4). The assessment of the impact of atmospheric conditions, namely, oxygen content and relative humidity, completes the examination of solar-cell relevant conditions (section 9.5).

9.1 Review of Studies on Compositional Stability of Mixed Perovskites

The first report of compositional instabilities due to segregation of the halides has been presented by Hoke *et al.* for $MAPb(I_{1-x}Br_x)_3$.[335] For bromine ratios $0.2 < x < 1$, they observed an additional red-shifted contribution in photoluminescence spectra with strongly increasing

[1] hereinafter referred to as $(Cs,FA,MA)Pb(I,Br)_3$ and abbreviated as Cs0.05 in chapters 7 and 8

intensity under 1 sun illumination in less than one minute which is attributed to the formation of iodine-rich and bromine-rich domains.[335] Because of the smaller energy gap for iodine-rich domains, the photoluminescence is dominated by the evolving red-shifted peak arising from these areas. The splitting of perovskite reflexes in X-ray diffractograms and increasing sub-bandgap absorption under illumination provide further evidence.[335] Surprisingly, the observed changes are fully reversible after a few minutes in the dark.[335]

On the basis of these surprising observations, various studies revealed further insights. Iodine vacancies are found to play a key role influencing the segregation process[335,464–466] and spatially-resolved measurements identified the grain boundaries as origin of low-energy emission from iodine-rich phases.[467,468] With respect to the performance of solar cells, iodine-rich domains can act as additional recombination centers (due to the smaller energy gap) that inhibit current flow in the device or reduce open-circuit voltage and fill factor.[468–471] However, the extent of performance deterioration varies significantly between different studies which is presumably caused by different crystal quality and device fabrication as well as measurement procedure (e.g., at maximum power-point or open-circuit conditions[472]).

Different models have been proposed to explain the origin of photo-induced halide segregation in mixed perovskites. Nevertheless, none of them can fully account for all reported observations. Calculations of the Helmholtz free energy suggest the formation of local minima at iodine-rich and bromine rich composition, respectively, under illumination which favors segregation.[413,473] However, this model does not provide any driving force for the demixing in the dark.

Bischak *et al.* proposed the formation of polarons due to large electron–phonon coupling in perovskites.[474] The stabilization of polarons at iodine-rich areas of initially present compositional fluctuations due to carrier localization (as a consequence of the smaller bandgap) and resulting lattice strain are assumed to cause accumulation of further iodide and, thereby, growth of iodine-rich domains.[413,466,474] Accordingly, homogeneous films with reduced halide fluctuations show less segregation.[413,474]

Further, a kinetic model developed by Draguta *et al.* identifies the reduced bandgap energy of the iodine-rich composition as driving force for the segregation by modifying the (initially nearly zero) free energy under illumination.[466,475] The extent of segregation is then also dependent on the diffusion length of charge carriers to reach iodine-rich domains. Thereby, the inhibited segregation in nanocrystals (with small diffusion length) up to a critical size of about 45 nm can be explained.[466,475,476] However, the proposed model cannot rationalize the reduced segregation in high-quality films with less grain boundaries and large charge-carrier diffusion lengths.[466,475]

Barker *et al.* focused on the role of iodine vacancies and found the low-bandgap iodine-rich phase at the illuminated side of the perovskite film.[465] Based on these spectroscopic results, they proposed a strong gradient in photogeneration of charge carriers across the film thickness due to an illumination gradient in the strongly absorbing perovskite and suggested a resulting inhomogeneous defect-assisted halide migration.[465]

Despite the still missing universal model for the segregation process, different strategies have been developed to reduce or suppress compositional instabilities. Other approaches rely on the significant role of defects, especially iodine vacancies, and apply, e.g., additional iodide treatment,[335,471] passivation of surfaces and grain boundaries using potassium doping,[174] or surface treatment of the perovskite layer[477] to reduce halide segregation.

In any case, the most common strategy is the incorporation of additional organic or inorganic A-site cations, e.g., FA^+ or Cs^+ (see chapter 2.1.3).[140,143,336,337,478] According to Bischak *et al.*, the less polar FA^+ and Cs^+ compared to MA^+ reduce the formation of polarons resulting in less segregation.[474] Similarly, the multiple-cation lead mixed-halide compound $Cs_{0.05}(FA_{0.83}MA_{0.17})_{0.95}Pb(I_{0.83}Br_{0.17})_3$ (hereinafter abbreviated as $(Cs,FA,MA)Pb(I,Br)_3$) originally introduced by Saliba *et al.*[167] is claimed to be intrinsically stable against halide segregation and presents the most widely used compound in perovskite solar cells. Therefore, the following chapter presents an in-depth study of compositional stability issues focusing on these mixed compounds.

9.2 Reversible Shifts of the Bandgap Energy in Electroreflectance Spectra Under Illumination and Bias

This section introduces the observed phenomena of reversible bandgap instabilities in $(Cs,FA,MA)Pb(I,Br)_3$ and presents first indications of their origin based on investigations of compounds with varied anion and cation compositions. Main parts of the measurements in this section were achieved in collaboration with Pascal Rietz within his bachelor's thesis[383] and are published in [241].

9.2.1 Fundamental Phenomena

Repeated electroreflectance measurements of devices with $(Cs,FA,MA)Pb(I,Br)_3$ absorber layer under identical conditions yield surprising results: The recorded spectra clearly vary despite the fact that $(Cs,FA,MA)Pb(I,Br)_3$ is reported to be more stable compared to other compounds.[143,167,336,337] In particular, not only the amplitude of the electroreflectance signal changes over time but – more importantly – the whole spectrum shifts toward lower energies. This behavior signifies a shift of the resonance energy E_{res} of the perovskite which is linked to its bandgap energy.

Figure 9.1 presents an exemplary study of these phenomena. An electroreflectance spectrum was recorded every 15 minutes (with a measurement time of ~ 7 minutes) using a sinusoidal ac bias with $V_{ac} = 0.5$ V and $V_{dc} = 0$ V (see Fig. 9.1 (a), green to red curves, only one spectrum per hour is depicted for the sake of clarity). In between the individual measurements, the ac bias was kept on and the sample was illuminated with a 1 sun solar spectrum to simulate solar-cell relevant conditions. The shift of the normalized transformed modulus spectra $|\mathcal{M}(E)|$ toward lower photon energies in Fig. 9.1 (b) clearly confirms the decrease of the resonance energy E_{res} which is determined as energy position of the peak maximum.

Here, it should be noted that the resonance energy E_{res} is not identical to the bandgap energy E_{g} in the investigated perovskite compounds but shifted by (about) the exciton binding energy E_{B} as pointed out in chapter 7. Still, it is reasonable to assume that the rather small exciton binding energy E_{B} does not significantly vary and relative changes of E_{res} observed here are mainly

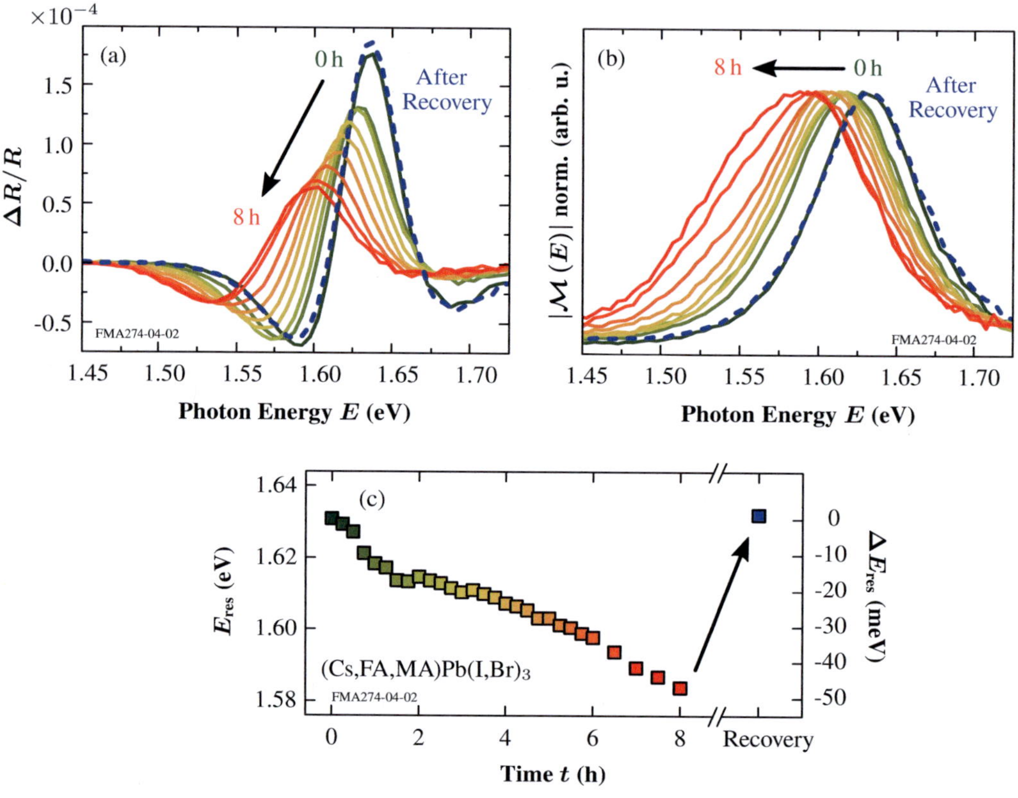

Figure 9.1: Reversible changes in (a) $\Delta R/R$ electroreflectance spectra and (b) corresponding transformed $|\mathcal{M}(E)|$ modulus spectra. Repeated measurements (green to red curves) have been performed every 15 min (with a measurement time of ~ 7 minutes) with 1 sun illumination and ac bias kept on in between two measurements (only one spectrum per hour depicted for clarity). In addition to a moderate decrease in intensity, the spectra clearly shift toward lower energies indicating a decrease of the resonance energy E_{res} and, thereby, the bandgap energy E_{g}. (c) E_{res} extracted from the energy position of the maximum in (b) drops by about 50 meV over 8 hours. Remarkably, the initial spectrum and, thereby, the resonance energy are fully recovered after storage in the dark in dry air for several days (dashed blue curves and blue square). Adapted with permission from [241]. Copyright 2018 American Chemical Society.

caused by changes of the bandgap energy E_g (which will become clearer with regard to the underlying mechanism revealed within the following sections). Accordingly, relative changes of E_{res} are also denoted as shifts of the bandgap energy E_g. In order to quantify the observed energy shifts, the peak position of the modulus spectrum $|\mathcal{M}(E)|$ is plotted over time in Fig. 9.1 (c) which shows a maximum bandgap decrease of about 50 meV after 8 hours of ac bias and additional 1 sun illumination (for about 50% of time as explained above).

Remarkably, the spectrum completely recovers its initial line shape after storage in the dark in dry air for several days (dashed blue curve)[2]. Accordingly, the original resonance energy E_{res} and, thereby, the bandgap energy E_g are also recovered as depicted in Fig. 9.1 (c). This exceptional reversibility of the obtained decrease in bandgap energy gives rise to the hypothesis of halide segregation as origin of the observed phenomena in (Cs,FA,MA)Pb(I,Br)$_3$ – despite the reported stability against segregation effects in these compounds.[143, 167, 336, 337] To further examine this assumption, studies of different compositions are presented in the following section.

9.2.2 Influence of Different Absorber Compositions

In order to identify the relevant cations or anions causing the observed bandgap instabilities, solar cells with different absorber compositions were investigated (see appendix F for details and power-conversion efficiencies). Starting from the full mixture of involved cations and anions, (Cs,FA,MA)Pb(I,Br)$_3$, successively ions were removed – first bromide followed by cesium and formamidinium which finally results in standard MAPbI$_3$. The observed resonance energies E_{res} over time (again, under illumination and applied bias) are summarized in Fig. 9.2 (a).

The overall dependence of the bandgap energy on the composition of the different compounds is in accordance with previous reports: Starting from MAPbI$_3$ (red squares), incorporation of formamidinium reduces the bandgap energy as observable for FA$_{0.83}$MA$_{0.17}$PbI$_3$ (abbreviated as (FA,MA)PbI$_3$, green squares). Additional cesium and bromine both lead to an increase of the optical gap, visible for both Cs$_{0.05}$(FA$_{0.83}$MA$_{0.17}$)$_{0.95}$PbI$_3$ ((Cs,FA,MA)PbI$_3$, yellow diamonds) and (Cs,FA,MA)Pb(I,Br)$_3$ (blue diamonds). In the latter case, the influence of FA as well as Cs and Br seem to roughly cancel out each other. Nevertheless, the most striking difference is clearly represented by the fact that only (Cs,FA,MA)Pb(I,Br)$_3$ exhibits a decrease in resonance energy E_{res} over time while no changes of E_{res} are observable for all other compositions. This strongly emphasizes the role of iodine and bromine with regard to the observed phenomena and further supports the conclusion of photo- and voltage-induced halide segregation. It should be noted here that a small irreversible shift of about 5 meV is detectable in (Cs,FA,MA)PbI$_3$ under high relative-humidity conditions which could be caused by formation of a Cs-rich secondary phase.[172] Similarly, a small irreversible contribution is also observed in (Cs,FA,MA)Pb(I,Br)$_3$ (see section 9.4.1).

[2] In this case, the sample was measured again after 14 days but was probably recovered after three to four days judging from similar measurements

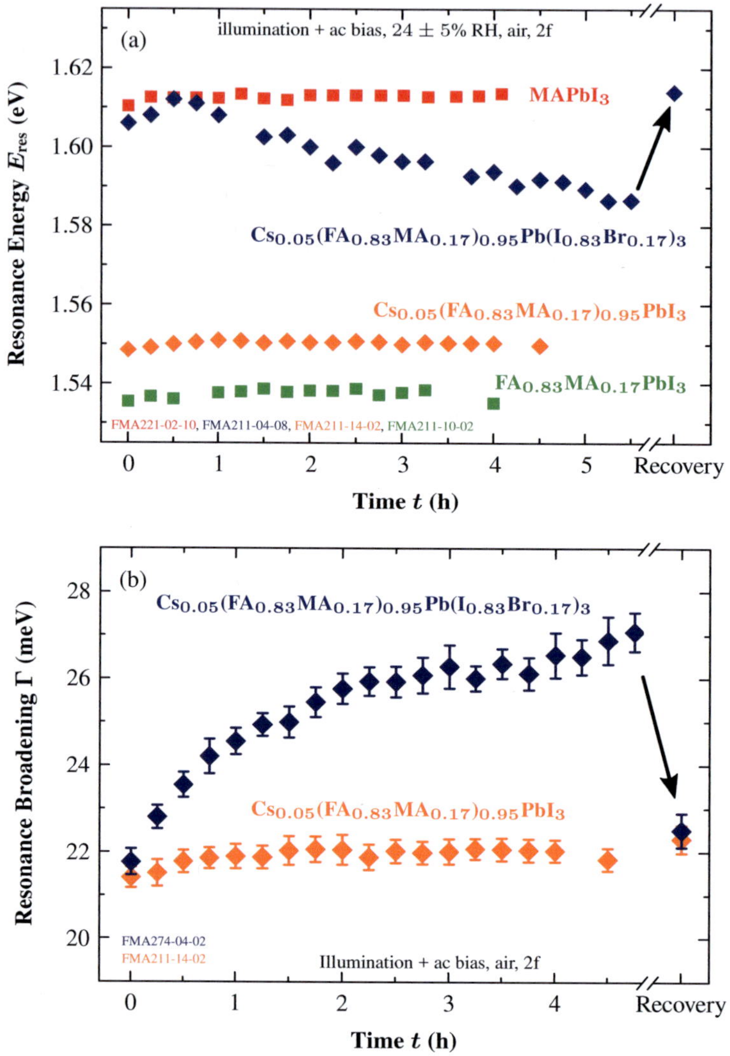

Figure 9.2: (a) Resonance energies E_{res} over time for different absorber compositions under illumination and bias. Clearly, only for (Cs,FA,MA)Pb(I,Br)$_3$ (blue diamonds) containing iodine and bromine, a reversible decrease of E_{res} is observable supporting the hypothesis of halide segregation as origin of the observed instabilities. (b) The resonance broadening Γ extracted from FDFF fits of measured electroreflectance spectra exhibits a clear increase over time (under illumination and applied bias) in case of (Cs,FA,MA)Pb(I,Br)$_3$ (blue diamonds). This indicates an increased inhomogeneity of the perovskite absorber layer which is in accordance with the assumption of halide segregation. After sufficient relaxation in the dark, the initial value for Γ is recovered. No similar behavior is observable in (Cs,FA,MA)PbI$_3$ without any bromine (yellow diamonds). Error bars denote fit uncertainties only. Reprinted with permission from [241]. Copyright 2018 American Chemical Society.

Additionally, not only the energy position of the resonance but also its (spectral) broadening can provide further information about the underlying mechanism. Qualitatively, an increase of the resonance broadening under illumination and bias is already visible in the transformed $|\mathcal{M}(E)|$ spectra in Fig. 9.1 (b) by comparison of the larger shift of the low-energy side of the resonance peak compared to its high-energy flank. Figure 9.2 (b) presents a more quantitative evaluation in the form of the time-dependent evolution of the resonance broadening Γ extracted from FDFF fits of the measured electroreflectance spectra. In case of (Cs,FA,MA)Pb(I,Br)$_3$ (blue diamonds), the clearly visible increase of the resonance broadening Γ confirms the trend in Fig. 9.1 (b). Similarly to the $|\mathcal{M}(E)|$ spectrum, the initial value of Γ is recovered after sufficient relaxation in the dark. Clearly, no similar behavior can be identified in case of (Cs,FA,MA)PbI$_3$ (yellow diamonds) which further confirms the absence of any photo- or voltage-induced changes in compounds without any bromine. The increasing resonance broadening over time in (Cs,FA,MA)Pb(I,Br)$_3$ indicates a rising inhomogeneity in the material which can be explained by segregation of the fully intermixed perovskite into iodine-rich and bromine-rich domains.

In summary, monitoring of photo- and voltage-induced changes in electroreflectance spectra of different perovskite compositions provides strong evidence for iodine- and bromine-related segregation within the absorber layer. The next step for further confirmation requires additional direct observations of structural or compositional alterations, e.g., obtained by X-ray diffraction.

9.3 Characterization of Photo- and Voltage-Induced Structural Changes

In order to verify the role of iodine- and bromine-related changes in the perovskite absorber for the observed photo- and voltage-induced bandgap instabilities, *in-situ X-ray diffraction* (XRD) is utilized to monitor changes in the crystal structure. Furthermore, *Time-of-Flight Secondary Ion Mass Spectrometry* (ToF-SIMS) provides spatially-resolved results on iodine and bromine content in complete solar cells.

9.3.1 Correlation of Bandgap-Energy Shifts to Structural Changes Using X-Ray Diffraction

X-ray diffraction patterns contain information about the crystal structure and lattice constant of a crystalline material. Therefore, specific materials and compositions can be identified by comparison with theoretically calculated patterns. In order to detect small compositional changes induced by illumination and applied bias, a conventional X-ray diffractometer has been equipped with a white-light LED solar simulator (with $\lesssim 1\,\mathrm{sun}$ due to tilted arrangement because of space limitations) and a function generator to provide a bias voltage.

These in-situ X-ray diffraction measurements presented within this chapter have been obtained at the *Ludwig-Maximiliam University* (LMU) in Munich in collaboration with Meltem F. Aygüler and are published in [241]. See appendix E.1 for further experimental details.

Figures 9.3 (a) and (b) present X-ray diffractograms obtained before (black curve) and after four hours of illumination and bias (red curve, similar as above ~ 1 sun and sinusoidal ac bias with $V_{ac} = 0.5$ V and $V_{dc} = 0$ V). An additional pattern was recorded after 7 days of relaxation in the dark (blue curve). The reflex at $26.5°$ can be assigned to the FTO layer which does not change during the four-hour treatment (and can be used as reference for the relaxation pattern). Both perovskite reflexes at $24.6°$ and $28.4°$ exhibit an asymmetric broadening after illumination and bias which is more apparent in the enlargement in Fig. 9.3 (b). The shoulder at smaller angles ($\sim 28.3°$) is attributed to an iodine-rich phase because the larger effective radius of iodide compared to bromide results in a larger lattice constant and, thereby, a smaller diffraction angle compared to the mixed iodine–bromine phase. Accordingly, the shoulder-like feature at larger angles ($\sim 28.5°$) probably stems from a bromine-rich phase. In addition, similarly to the optical spectra, the initial diffractogram is (nearly) recovered after relaxation in the dark.

Additionally, the characteristic (100) perovskite reflex at $14.1°$ was monitored in situ under illumination and applied bias for five hours (see Fig. 9.3 (c), green to red curves). The observed asymmetric broadening of the (normalized) diffraction patterns toward smaller angles presumably corresponds to a similar evolution of the peak as in Fig. 9.3 (a) and (b) but here it cannot be clearly resolved. Still, the *Full-Width Half-Maximum* (FWHM) can be utilized as a quantitative measure and is depicted over time in Fig. 9.3 (d) (colors correspond to the diffraction patterns in Fig. 9.3 (c)). The increasing FWHM is in accordance with the results obtained by electroreflectance spectroscopy in Fig. 9.2 (b) and is attributed to the rising inhomogeneity of the perovskite due to progressing segregation into iodine-rich and bromine-rich domains. After termination of illumination and applied bias, the perovskite reflex relaxes towards its initial shape over the course of days which is visible in both the diffraction patterns (dashed purple and blue curve) in Fig. 9.3 (c) and the FWHM in Fig. 9.3 (d).

The analysis of in-situ X-ray diffraction measurements confirms reversible changes in the crystal structure of the perovskite absorber in complete solar cells induced by illumination and ac bias. These results are consistent with photo- and voltage-induced reversible bandgap instabilities observed in electroreflectance spectra in terms of increasing broadening and reversibility and can be attributed to a segregation of the initially intermixed perovskite into iodine-rich and bromine-rich domains.

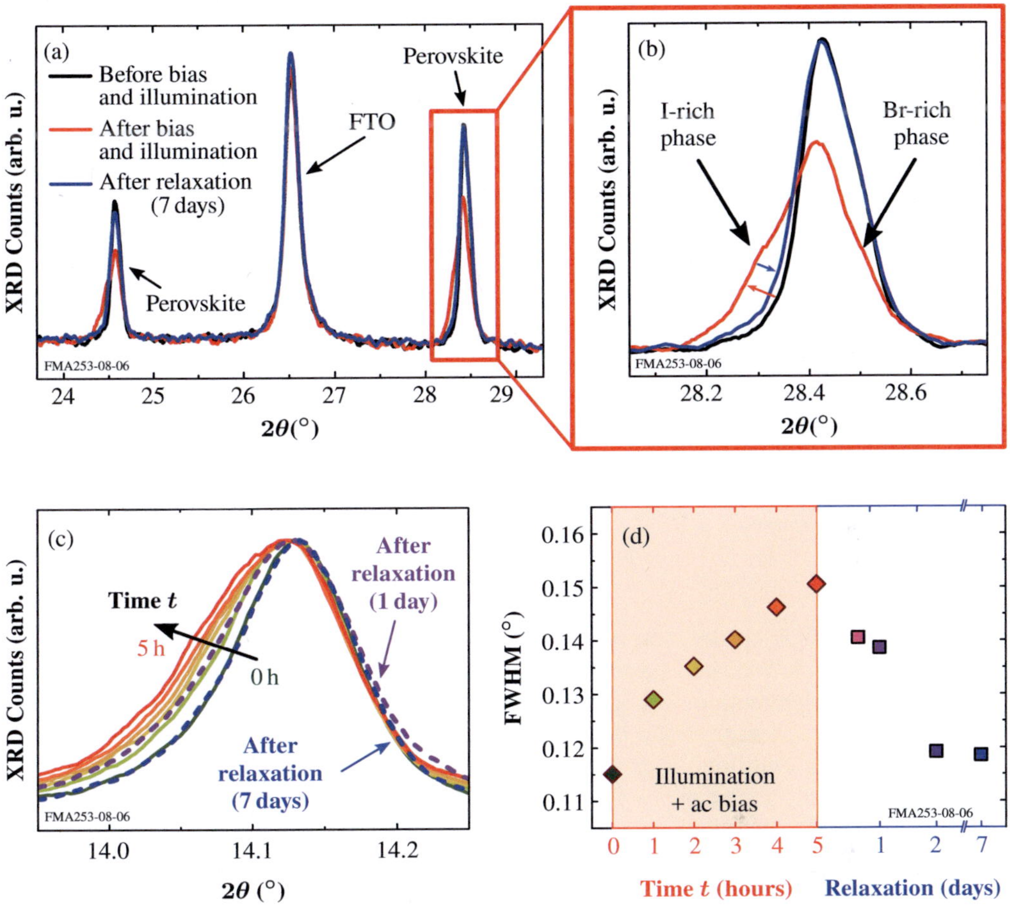

Figure 9.3: (a) X-ray diffractogram before (black curve) and after four hours under illumination and applied ac bias (red curve). The initial perovskite reflex at 28.4° (and 24.6°) starts to split into two contributions which are visible as shoulder-like features at smaller ($\sim 28.3°$) and larger angles ($\sim 28.5°$) in the enlargement in (b). They can be attributed to an iodine-rich and bromine-rich phase, respectively, due to the different lattice constants and corresponding diffraction angles. After one week relaxation in the dark, the initial diffraction pattern is (nearly) recovered. (c) Monitoring of the perovskite reflex at 14.1° under white-light illumination and applied ac bias for 5 h (green to red curves). The apparent asymmetric broadening on the small-angle side is attributed to increasing inhomogeneity caused be progressing halide segregation. The diffraction pattern (nearly) recovers its initial shape after storage of several days in the dark (dashed purple and blue curve). (d) The FWHM of the perovskite reflex in (c) is utilized as quantitative measure and increases accordingly under illumination and bias. In the dark without applied voltage, the FWHM relaxes towards its initial value after several days (colors correspond to the curves in (c) but with two additional data points for relaxation). Reprinted with permission from [241]. Copyright 2018 American Chemical Society.

9.3.2 Time-of-Flight Secondary Ion Mass Spectrometry (ToF-SIMS)

The mass analysis of secondary ions ejected from the surface of a sample of interest due to sputtering with so-called primary (or probing) ions is a useful tool for determination of the elemental composition (see appendix E.2 for further details and experimental setup).[353] In particular, depth profiling is possible over several micrometers[353] and yields, e.g., the spatial distribution of different ions in a perovskite solar-cell stack. Thereby, additional information about the photo- and voltage-induced changes in halide composition within the perovskite absorber layer can be gathered.

Figure 9.4 depicts depth profiles of two samples together with approximate positions of the different layers in the layer stack.[3] One sample was treated similarly to the previous electroreflectance and X-ray diffraction measurements by illumination with a solar spectrum (1 sun) and applied bias (here: reverse bias condition with $V_{dc} = -0.5\,\mathrm{V}$), in this case, for two times five hours (bright curves). After six days of relaxation in the dark, an additional measurement was performed (dashed lines). A second untreated control samples is used for comparison (dark curves). All measurements were designed together with and have been performed by Dr. Jonas Hanisch at the *Zentrum für Sonnenenergie- und Wasserstoff-Forschung Baden-Württemberg* (ZSW) in Stuttgart.

Both the iodide (blue curves) and the bromide (red curves) signal exhibit an increase in the spiro-OMeTAD hole-transport layer in case of the treated sample (blue and red arrow) which is a result of the high ion mobility in the device. While the iodide signals remains at a higher level after six days storage in the dark the changes in case of bromide are partly reversible (green arrow). In addition, after illumination and bias, a local minimum of the bromide signal appears in the center of the perovskite absorber layer which evens out after relaxation. Thereby, ToF-SIMS investigations confirm the presence of reversible changes of the bromine distribution within the perovskite absorber and – more surprisingly – the spiro-OMeTAD hole-transport layer. The detected reversible redistribution of bromide ions presumably causes the observed reversible changes in X-ray diffractograms and electroreflectance spectra described above. The irreversible contributions of iodide and, partly, bromide in the spiro-OMeTAD layer are attributed to simultaneous aging under ambient conditions. This is supported by additional measurements of completely fresh samples which show significantly reduced iodide signals in the spiro-OMeTAD layer (more than one order of magnitude). Similarly, the increased contribution of gold within the perovskite (yellow arrow) in case of the illuminated and biased sample can be a result of gold diffusion into the absorber layer (and adjacent layers) causing irreversible degradation of the device as previously reported.[114, 131]

The apparent changes within the spiro-OMeTAD layer raise questions about the influence of different hole-transport materials and their interface to the perovskite layer on the halide segregation. While so far most studies focused on bare perovskite films only few reports of

[3] The positions of the different layers are estimated from additional signals not displayed here such as TiO^- or PbI_3^-. Their apparent width is based on the required sputter time and depends on the material properties and sputter source.

interfacial properties and their modification are published and show the potential to influence and reduce halide segregation.[174,477] First investigations of different hole-transport layers using electroreflectance spectroscopy indeed indicate variations in segregation rates and extent.[479] However, further systematic examinations are required for detailed understanding.

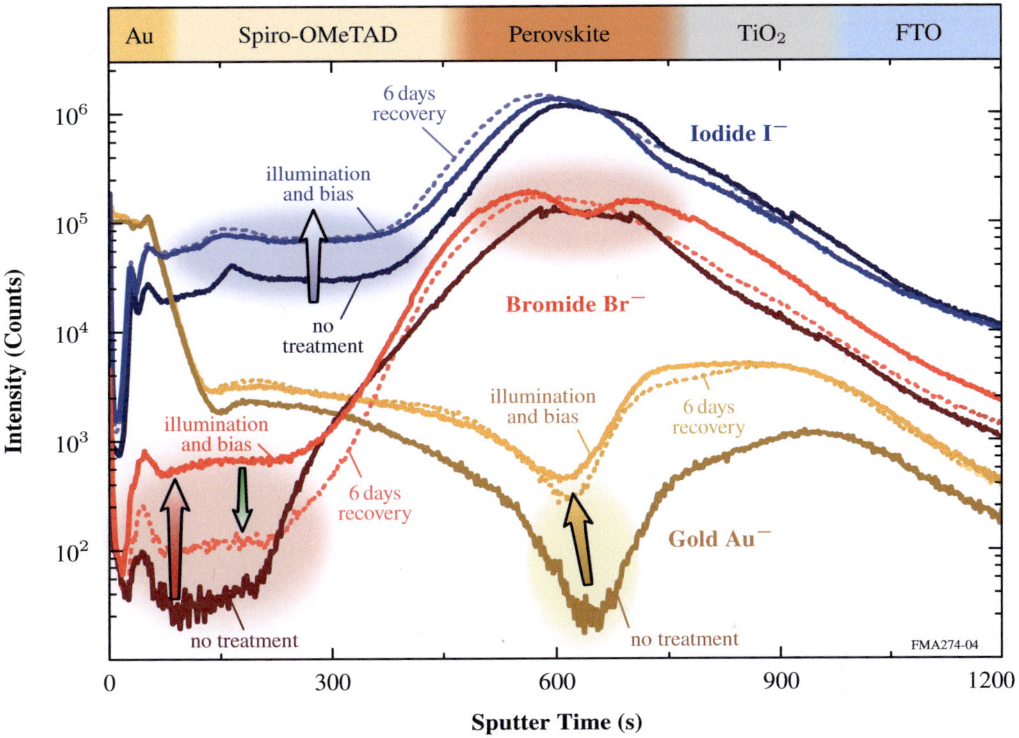

Figure 9.4: Comparison of *Time-of-Flight Secondary Ion Mass Spectrometry* (ToF-SIMS) results of a sample with applied reverse bias ($V_{dc} = -0.5$ V) and 1 sun illumination for 2×5 h (bright solid lines) and a control sample without treatment (dark solid lines). Additional results of the treated sample after six days of recovery in the dark are depicted (dashed lines). Iodide (blue curves) and bromide (red curves) signal show a clearly observable increase in the spiro-OMeTAD layer for the treated sample which is partly reversible in case of bromide. Additionally, a local minimum in the bromide signal appears in the perovskite absorber which evens out after relaxation. The reversible redistribution of bromide is assumed to be responsible for the observed reversible bandgap-energy shifts. Irreversible alterations are attributed to simultaneous aging of the sample in air. Furthermore, the increased gold content (yellow curves) in the absorber of the biased and illuminated sample can indicate previously reported diffusion of gold atoms from the contact layer which degrades the perovskite absorber.[114,131] Measurements have been performed by Jonas Hanisch at the *Zentrum für Sonnenenergie- und Wasserstoff-Forschung Baden-Württemberg* (ZSW) in Stuttgart.

9.4 The Role of Illumination and Voltage for Induced Bandgap Instabilities

The next step after determination of the origin of the observed changes in resonance energy E_{res} is a more detailed examination of external stimuli of halide segregation – in this case, light and applied bias. This is even more important in the context of perovskite solar cells since illumination and voltage represent intrinsic solar-cell working conditions. In order to enable a better quantitative comparison of the impact of different external influences, a modified measurement protocol has been developed which is introduced in the following and applied for the examination of reversibility and photo- and voltage-induced effects. Main parts of the measurements in this and the following section were achieved in collaboration with Pascal Rietz within his bachelor's thesis[383] and are published in [241].

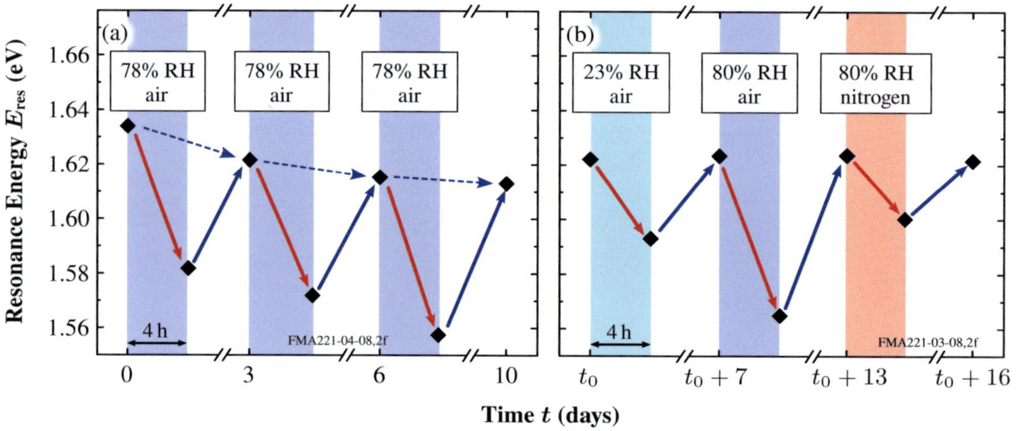

Figure 9.5: Measurement protocol used for better quantitative comparison of the impact caused by different external influences. After initial determination of the resonance energy E_{res} a defined influence is applied for 4 h (red arrows and colored areas), e.g., illumination and/or applied bias, or different atmospheric conditions, followed by a second measurement of E_{res} for comparison. Prior to the next investigation, the sample is stored in dry air in the dark for at least three days for recovery (blue arrows, note the number of days on the time axis). (a) The first few measurements of fresh sample show a superimposed small irreversible decrease in resonance energy (see dashed blue arrows) which is taken into account in comparisons of different external influences. (b) Different atmospheric conditions (or external stimuli) result in a different extent of resonance-energy decrease. However, after sufficient relaxation in the dark, the initial E_{res} is recovered. Reprinted with permission from [241]. Copyright 2018 American Chemical Society.

9.4.1 Reversibility of Bandgap-Energy Shifts

The modified measurement protocol used for the following investigations is visualized in Fig. 9.5. First, the resonance energy of the perovskite is determined by recording an electroreflectance spectrum and performing the modulus transformation. Thereafter, a defined influence is applied for a duration of four hours (red arrows and colored areas), e.g., illumination and/or applied bias, or different surrounding atmospheric conditions (depicted in Fig. 9.5 (b), see further discussion in section 9.5). Finally, the resonance energy is measured again for comparison with the initial value. Prior to the subsequent measurement, the sample is stored in dry air in the dark for at least three days (blue arrows) to ensure complete recovery (Note that the time axis is given in days but the colored areas denote only four hours.).

Figure 9.5 (b) shows the reversibility of the bandgap-energy shifts for different surrounding atmospheres which is discussed in section 9.5 in detail. However, for the very first measurements of fresh samples a small irreversible shift (dashed blue arrows) is superimposed to the reversible effects (see Fig. 9.5 (a)). After a few measurements the resonance energy of the recovered solar cells stabilizes at 1.61 eV − 1.62 eV. Since a small initial shift is also observed in (Cs,FA,MA)PbI_3 but not (FA,MA)PbI_3 and $MAPbI_3$, this could be related to an irreversible formation of a Cs-rich secondary phase.[172] More importantly, for comparison of results under different conditions or stimuli, resonance energies E_{res} were always determined after stabilization to ensure reliable results. Moreover, repeated measurements with varied order of different external influences were performed to avoid any influence of aging or "memory" effect.

9.4.2 Impact of Light and Applied Bias on Bandgap Instabilities

Using the measurement protocol introduced in the previous section, repeated experiments were conducted with either 1 sun illumination, applied ac bias, or both combined in order to disentangle the influence of the individual stimuli. A representative example is depicted in Fig. 9.6.

As expected, no shift of the resonance energy E_{res} occurs for the reference experiment with neither additional illumination nor applied bias. This also verifies the negligible influence of the ac bias and illumination[4] required during the electroreflectance measurement (for ∼ 7 minutes). In case of 1 sun illumination only, a small decrease of E_{res} of less than 10 meV is observed which, in a sense, confirms the enhanced stability of multiple-cation compounds compared to, e.g., $MAPb(I_{1-x}Br_x)_3$ from previous reports.[167,336,337,478] However, in order to fully characterize solar-cell stability under relevant conditions, electric fields have to be taken into account, in this study by an applied bias. The respective experiment with applied voltage but without additional illumination shows that halide segregation can be induced even in the dark.

[4] The intensity of the tunable light source is estimated to be in the order of $\mu W/mm^2$ and, thus, significantly lower than $1\,mW/mm^2$ in case of 1 sun illumination

In case of (Cs,FA,MA)Pb(I,Br)$_3$, the decrease in resonance energy of about 20 meV actually exceeds the one under illumination only. Nevertheless, the by far largest shifts of about 50 meV or even above occur in case of both 1 sun illumination and applied bias.

Although various models to describe and explain halide segregation in mixed perovskites propose different driving forces, e.g., strain induced by polaron creation and stabilization[474] or a gradient in trap formation due to inhomogeneous light intensity[465] (see section 9.1), the common feature is the migration and redistribution of halide ions. Therefore, in any case, the additional applied voltage supports these mechanisms, e.g., in terms of additional activation energy or microstrain, resulting in increased ion migration and, thereby, halide segregation. Further correlation of the separate influence of illumination or voltage and structural characterization such as X-ray diffraction or ToF-SIMS could help to reveal the underlying processes in more detail.

In summary, the presented results emphasize the importance to include electric fields (here as applied bias) in stability considerations of perovskite solar cells and, in particular, in case of halide segregation. Moreover, bandgap instabilities in (Cs,FA,MA)Pb(I,Br)$_3$ occur most strongly under combined 1 sun illumination *and* applied voltage. Since both light and electric fields are intrinsic to solar-cell operation conditions, these effects cannot be avoided completely.

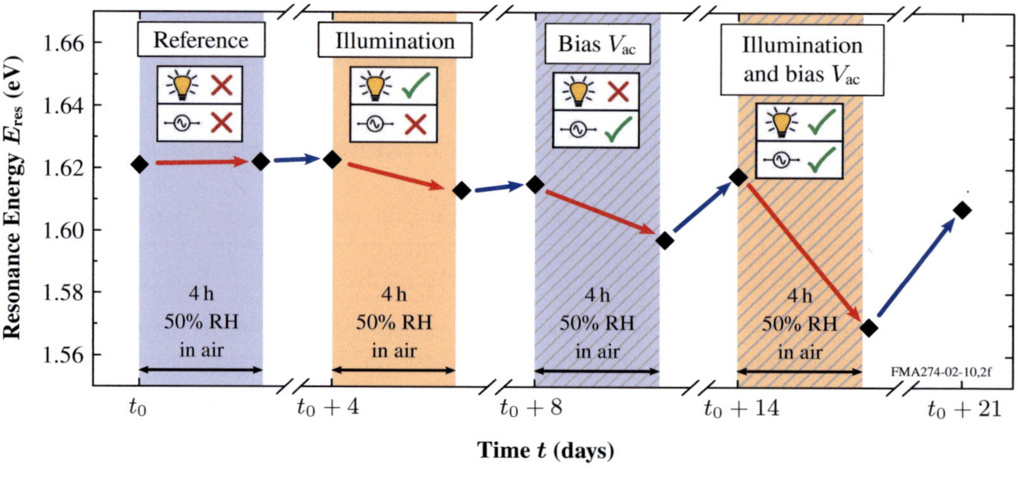

Figure 9.6: The influence of illumination and applied bias on bandgap instabilities. As expected, the reference experiment with neither illumination nor voltage does not show any change. In case of illumination only, a small shift of less than 10 meV supports the previously reported enhanced robustness against halide segregation of multiple-cation perovskites compared to, e.g., MAPb(I$_{1-x}$Br$_x$)$_3$. The reversible decrease of about 20 meV with only applied bias demonstrates that halide segregation can even be induced in the dark. Still, the strongest change of about 50 meV occurs under combined illumination and applied voltage. Reprinted with permission from [241]. Copyright 2018 American Chemical Society.

9.4.3 Effects Under DC Bias Conditions

In order to complete the investigations of the role of illumination and voltage on halide segregation in (Cs,FA,MA)Pb(I,Br)$_3$, effects of an applied dc voltage (instead of ac bias as above) in forward or reverse bias were studied providing better emulation of realistic solar-cell conditions. For both polarities of the applied dc bias (forward and reverse) qualitatively similar changes in $|\mathcal{M}(E)|$ modulus spectra were observed as presented in Fig. 9.7. For instance, in case of a dc forward bias of $V_{dc} = 0.5$ V (and 1 sun illumination), the $|\mathcal{M}(E)|$ spectra exhibit a transition from the initial resonance peak (green curve) at 1.63 eV towards a second low-energy contribution evolving at 1.55 eV after about $6 - 8$ hours (orange curves).[5] Finally, after 12 hours (red curve), the initial resonance peak is strongly decreased in intensity and is only visible as an additional shoulder-like feature. To reduce the potential influence of the ac bias applied during electroreflectance measurements,[5] additional experiments were performed consisting of only one recorded spectrum before and after four hours of applied dc bias and illumination.

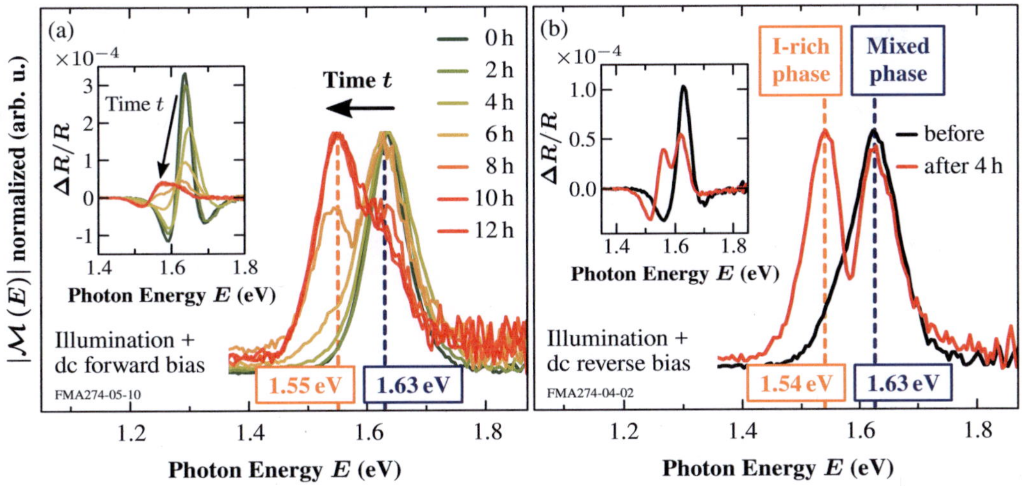

Figure 9.7: Halide segregation effects under illumination and dc bias conditions ($V_{dc} = \pm 0.5$ V). (a) Temporal evolution (green to red curves) of $|\mathcal{M}(E)|$ modulus and $\Delta R/R$ spectra (see inset) in case of a dc forward bias. While the initial resonance peak at 1.63 eV (nearly) vanishes a second low-energy contribution at 1.55 eV arises after $6 - 8$ hours. (b) Electroreflectance measurements before (black curve) and after (red curve) four hours under illumination and dc reverse bias show a comparable behavior and exhibit a clear coexistence of both contributions. The initial peak at 1.63 eV (blue dashed line) and the second evolving contribution at 1.54 eV – 1.55 eV (yellow dashed line) are – for both forward and reverse bias – assigned to an intermixed iodine–bromine phase and an iodine-rich phase, respectively. Reprinted with permission from [241]. Copyright 2018 American Chemical Society.

[5] Electroreflectance spectra are recorded every 30 minutes. Since an ac bias is required during the measurement duration of ~7 minutes, (forward or reverse) dc bias and 1 sun illumination are only applied for about 75% of the time.

Figure 9.7 (b) displays an example in case of a dc reverse bias of $V_{dc} = -0.5\,\mathrm{V}$. After four hours of combined illumination and dc bias (red curve), the coexistence of both the initial resonance peak and the additional low-energy contribution is clearly visible in $\Delta R/R$ and the transformed $|\mathcal{M}(E)|$ modulus spectrum.

The energy position of the low-energy peak at about 1.55 eV agrees well with the resonance energy of the bromine-free (Cs,FA,MA)PbI_3 perovskite modification (for both forward and reverse bias). Therefore, it is assigned to an evolving iodine-rich phase because of the migration of bromide – probably, at least partly, into the spiro-OMeTAD. Consequently, the initial peak at 1.63 eV is attributed to an intermixed iodine–bromine phase. The coexistence of both contributions in electroreflectance spectra indicates the formation of domains with different bromine content. However, the behavior under dc bias conditions including the formation of an additional low-energy resonance feature and the coexistence of two distinct contributions differs apparently from the increasing broadening and continuous shift of the resonance peak in case of applied ac bias. The latter can be attributed to the formation of an increasing number of small iodine-rich domains within the perovskite absorber which are presumably located at grain boundaries according to literature.[467,468] Consequently, only an "effective" bandgap energy of the perovskite is observed in electroreflectance spectra which decreases with the number of iodine-rich domains and broadens due to the increased inhomogeneity. Similarly, in case of the unidirectional nature of the electric field for an applied dc bias, the formation of larger domains or accumulation of small domains – possibly at ETL and HTL interfaces – is hypothesized which lead to resolvable contributions of both iodine-rich and intermixed phase. Again, the spiro-OMeTAD layer plays an important role and diffusion of bromide into it could account for missing bromine-rich signatures in electroreflectance spectra. Further investigations and correlation with spatially resolved determination of iodine and bromine content could further reveal the exact underlying mechanism.

9.5 Influence of Atmospheric Conditions on Bandgap Instabilities

In addition to illumination and voltage, the surrounding atmosphere has a significant influence on the halide segregation in the absorber and, thereby, the observed bandgap-energy shifts. Using the home-built setup presented in chapter 6.1 both surrounding gas and relative humidity can be controlled and their impact on electroreflectance spectra and corresponding resonance energies E_{res} can be studied systematically.

9.5.1 Changes in Electroreflectance Spectra Under High Relative Humidity Conditions

Prior to a systematic quantification of the influence of different atmospheric conditions on bandgap-energy shifts, it is reasonable to scrutinize complete electroreflectance spectra.

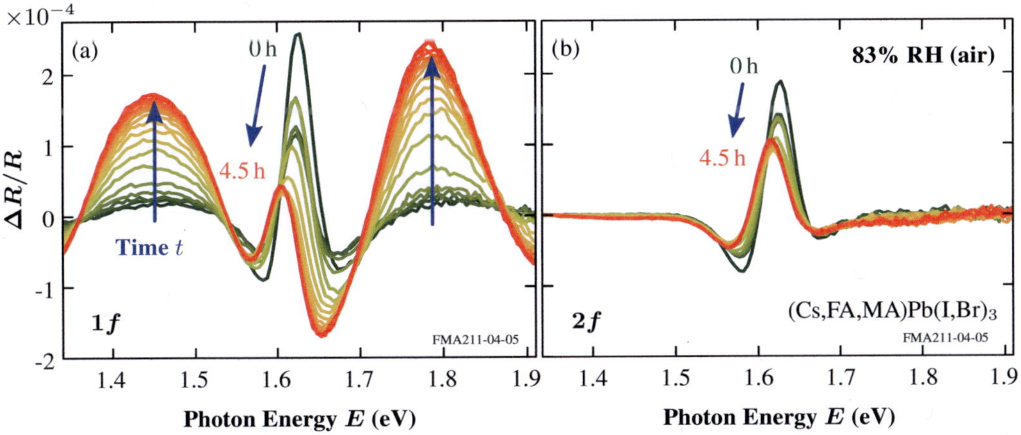

Figure 9.8: Time evolution of electroreflectance spectra detected at (a) fundamental frequency $1f$ and (b) second-harmonic frequency $2f$ at 83% relative humidity in air (recorded every 15 min). No additional voltage or 1 sun illumination are applied except for the ac bias (and low-intensity illumination of the tunable light source) during measurement which is responsible for the small shift of less than 10 meV. After approximately one hour, strong additional contributions around 1.45 eV and 1.79 eV occur and rise until saturation after about 3 hours. Spectra detected at second-harmonic frequency $2f$ do not exhibit similar changes and only show a small shift related to halide segregation as before. Adapted from [383].

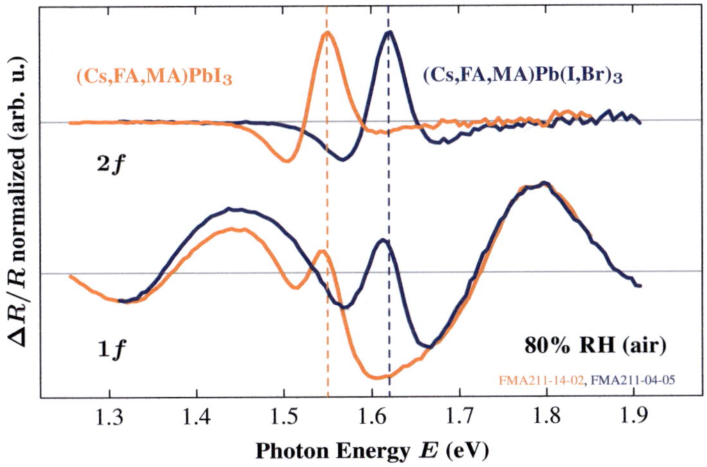

Figure 9.9: Electroreflectance spectra detected at second-harmonic ($2f$) and fundamental ($1f$) frequency for (Cs,FA,MA)Pb(I,Br)$_3$ (blue curves) and (Cs,FA,MA)PbI$_3$ (yellow curves) at 80% relative humidity in air. While spectra detected at $2f$ exhibit a distinct resonance feature enabling reliable determination of the resonance energy E_{res} strong additional contributions appear in case of $1f$ detection. Interestingly, these additional features occur at identical energies despite the different absorber composition. Data published in [383].

In particular, for high relative humidity ($\sim 80\%$) strong changes appear. Figure 9.8 shows the time evolution (green to red curves) of $\Delta R/R$ electroreflectance spectra detected at fundamental frequency $1f$ in (a) and second-harmonic frequency $2f$ in (b) at 83% relative humidity in air. Spectra were recorded every 15 minutes without additional 1 sun illumination or applied voltage except for the ac bias and low-intensity illumination of the tunable light source which are needed for electroreflectance measurements. The influence of the measurements causes only a small shift of less than 10 meV accompanied by a minor decrease in amplitude (see in Fig. 9.8 (b)). However, clearly visible additional contributions around 1.45 eV and 1.79 eV appear in case of $1f$ detection (see Fig. 9.8 (a)) after approximately one hour and saturate after about three hours. Interestingly, these effects are only observable at high relative humidity and show only minor contributions for 25% and 55% relative humidity.[383] The observed additional contributions for $1f$ are probably related to linear electro-optical effects which are suppressed in second-harmonic detection.[341]

Furthermore, the comparison with electroreflectance spectra for (Cs,FA,MA)PbI_3 indicates similar additional features for $1f$ detection at identical energy positions despite the clearly different resonance energies (see Fig. 9.9). Moreover, the contribution around 1.79 eV is even present in $MAPbI_3$ and (FA,MA)PbI_3.[383] Therefore, the origin of these effect could be related either to the identical layer stack of all samples including electron- and hole-transport layer or a mainly composition-independent process in the perovskite absorber. The latter might be linked to the reversible formation of a hydrated perovskite as observed in $MAPbI_3$.[325]

The time scales of $30-60$ minutes for hydration and about 10 minutes for the reverse process are consistent with the alterations of electroreflectance spectra.[325] In any case, the spectra detected at second-harmonic frequency $2f$ show distinct and reliable resonance features even for high relative humidity which allow for a precise determination of the resonance energy E_{res} and were used throughout this work. However, the high accuracy and reliability are even more important in case of the small shifts of the resonance energy induced by halide segregation effects.

9.5.2 Impact of Oxygen Content and Relative Humidity on Bandgap-Energy Shifts

Although halide segregation effects in (Cs,FA,MA)Pb(I,Br)$_3$ are clearly induced by illumination and bias as demonstrated in section 9.4.2, the extent of the resulting reversible shifts of the resonance energy E_{res} is also significantly influenced by additional factors, in particular, the surrounding atmosphere. Therefore, the impact of high and low relative humidity, i.e., $(23 \pm 5)\,\%$ and $(81 \pm 7)\,\%$, respectively, and oxygen content (pure nitrogen, air, or pure oxygen) was systematically studied. The resulting shifts ΔE_{res} of the resonance energy after four hours of 1 sun illumination and applied ac bias ($V_{ac} = 0.5\,V$, $V_{dc} = 0\,V$) are summarized in Fig. 9.10 (a). First, by focussing on air as surrounding gas atmosphere (blue boxes), an apparent increase of the bandgap-energy shifts from about 30 meV to (50 ± 13) meV accompanied by a larger spread of the values is observed with increasing relative humidity. This tendency is even enhanced in pure oxygen atmosphere (green boxes) yielding maximum shifts exceeding 70 meV for high relative humidity.[6] In contrast, no significant trend is observable in case of pure nitrogen atmosphere (red boxes).

In summary, a small bandgap decrease of about 20 meV to 30 meV occurs for low relative humidity independent of the surrounding oxygen content. However, the combined influence of both high relative humidity and high oxygen content strongly boosts the extent of the observed bandgap-energy shifts.[7] In addition, Figure 9.10 (b) presents the resonance energy E_{res} before and after the four hours of illumination and applied ac bias in case of high relative humidity. Starting at a comparable resonance energy of about 1.62 eV, the different extent of bandgap-energy shifts (presented in Fig. 9.10 (a)) leads to final values ranging from 1.60 eV to 1.55 eV with increasing oxygen content (pure nitrogen to air to pure oxygen).

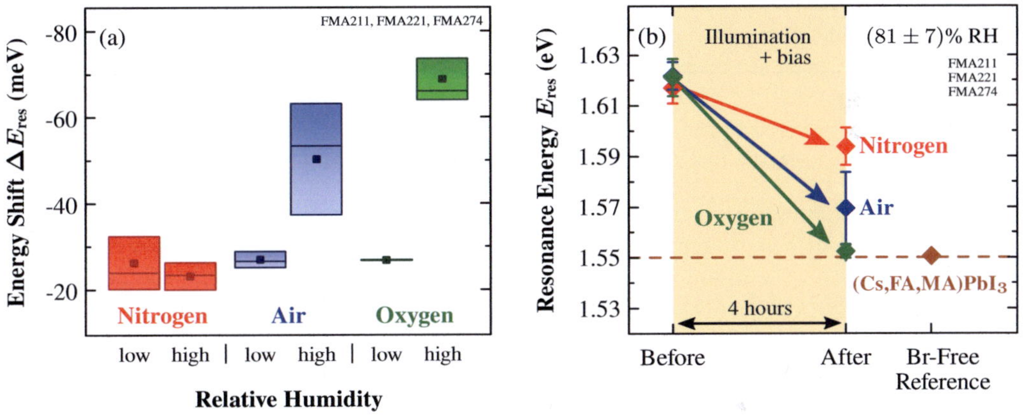

Figure 9.10: Influence of atmospheric conditions on bandgap instabilities in (Cs,FA,MA)Pb(I,Br)$_3$. (a) Resonance-energy shifts ΔE after 4 hours of 1 sun illumination and ac bias ($V_{ac} = 0.5$ V, $V_{dc} = 0$ V) for pure nitrogen (red), air (blue), or pure oxygen (green) and low ((23 ± 5) %) or high ((81 ± 7) %) relative humidity. A small shift between 20 meV and around 30 meV is observed for low relative humidity independent of the oxygen content. However, large values exceeding 70 meV occur only under the combined influence of both high relative humidity and high oxygen content. (b) Resonance energy E_{res} before and after 4 hours of illumination and bias in case of high relative humidity. For pure oxygen (green diamonds), the lowest final E_{res}, i.e., the largest shift, is obtained which corresponds well to the value of (Cs,FA,MA)PbI$_3$ without any bromine. This observation provides further evidence of strong iodide–bromide segregation in (Cs,FA,MA)Pb(I,Br)$_3$. (b) Adapted with permission from [241]. Copyright 2018 American Chemical Society.

[6] It should be noted here that less measurements have been performed under pure oxygen atmosphere and, unfortunately, only one in case of low relative humidity.

[7] The investigations of the role of illumination and bias in Figs. 9.6 and 9.7 have been performed at an intermediate relative humidity of 50%.

In particular, the latter value of 1.55 eV is the lowest one observed – even under most extreme conditions – and clearly matches the resonance energy of the (Cs,FA,MA)PbI_3 compound (brown diamond) containing no bromine at all. These observations provide further evidence for strong iodide–bromide segregation in (Cs,FA,MA)$Pb(I,Br)_3$ and are consistent with the evolving additional low-energy contribution in case of applied dc bias (see section 9.4.3).

The larger shifts of the resonance energy for high relative humidity can be attributed to an increased mobility of the halide ions.[480,481] The enhanced halide mobility supports faster segregation and, thereby, results in a larger shift of the effective bandgap energy. In particular in the context of a defect-assisted mechanism, this accelerated ion migration can be perceived as most relevant effect.[465,466] In addition, the formation of a monohydrated perovskite under high relative humidity[172,325] can reduce the homogeneity of the absorber layer and favor segregation as explained by the polaron-assisted model of Bischak *et al.*[413,474] In that sense, the observed changes in $\Delta R/R$ spectra detected at $1f$ under high relative humidity could be an indication for hydration of the perovskite (see section 9.5.1) which, in turn, might indirectly affect halide segregation. However, the exact mechanism remains unresolved and, in particular, the role of oxygen is still elusive. The present oxygen probably leads to the formation of O_2^- superoxides and, thereby, an increased number of defects.[328,330] This could explain an increased defect-assisted halide segregation but not the reversibility of the observed bandgap instabilities.

In summary, although the segregation into iodine-rich and bromine-rich domains in (Cs,FA,MA)$Pb(I,Br)_3$ is driven by illumination and electric fields which represent intrinsic solar-cell operation conditions, its extent could be reduced by effective encapsulation against oxygen and moisture as pointed out by the results in this section.

9.6 Summary

In this chapter, compositional instabilities in multiple-cation mixed-halide perovskite solar cells under illumination and applied voltage are thoroughly examined using electroreflectance spectroscopy. While stability is identified as one of the key challenges for the future of perovskite solar cells, unfortunately, mixed perovskite compounds with promising properties are prone to demixing, e.g., by photo-induced halide segregation in $MAPb(I_{1-x}Br_x)_3$.[335,413,466] Still, multiple-cation perovskites such as (Cs,FA,MA)$Pb(I,Br)_3$ representing the most prominent and relevant example in perovskite photovoltaics today are claimed to be stable against halide segregation under illumination.[143,167,336,337]

Nevertheless, by using electroreflectance as a very sensitive tool, a reversible decrease of the bandgap energy accompanied by a broadening of the resonance peak under 1 sun illumination and applied voltage was detected. A subsequent analysis of different perovskite compositions identifies bromine and iodine as the origin of these bandgap instabilities. This observation is confirmed by reversible changes in X-ray diffractograms recorded in-situ under illumination and bias which are assigned to the formation of iodine-rich and bromine-rich domains. Furthermore, a reversible redistribution of bromide in both the perovskite absorber and the spiro-OMeTAD hole-transport layer is observed in spatially-resolved ToF-SIMS results after illumination and applied voltage. Additional irreversible alterations are attributed to simultaneous aging under ambient conditions.

A detailed analysis of the role of light and voltage as driving forces for segregation reveals that the combined influence of both 1 sun illumination and applied ac bias results in the largest shifts of the resonance energy – clearly exceeding the individual contributions induced by light or bias only. Furthermore, under dc bias conditions, a coexistence of the initial resonance peak and an additional contribution at lower energies evolving after several hours of illumination and bias is observed. Thus, the formation of an increasing number of iodine-rich domains is proposed in the case of an ac bias resulting in a decreasing "effective" bandgap energy. The clearly resolvable low-energy contribution under dc bias conditions can then be explained by the formation or accumulation of larger iodine-rich domains. In addition, atmospheric conditions influence the extent of the bandgap instabilities. The rather small changes of about 20 meV – 30 meV in case of low relative humidity are strongly increased for both high relative humidity and high oxygen content exceeding 70 meV. These results could open up the possibility to reduce detrimental effects of instabilities by effective device encapsulation. Still, they cannot be avoided completely since illumination and voltage are intrinsic to the solar cell's operation conditions. Therefore, further development of alternative compositions towards single-halide and completely inorganic perovskites is essential for reliable long-term stability.

Chapter 10

Summary and Outlook

Summary and Conclusions of This Work

In recent years, thin-films solar cells based on organic–inorganic perovskites have attracted a great deal of attention which led to vibrant and wide-ranging research activities. As a consequence, remarkable achievements have been realized including an unprecedented rise of power-conversion efficiencies exceeding 24%.[10] Still, these impressive results are not in all cases accompanied by a profound understanding of underlying physical processes.

In this work, optical spectroscopy techniques were used to gain deeper insights into both excitonic effects influencing the fundamental working principle and practical stability considerations which are important for commercialization of this technology. In particular, electromodulation spectroscopy has been established on complete working perovskite solar cells enabling the precise and reliable determination of the energy of fundamental optical resonances. The application of this sensitive spectroscopic method to different use cases represents the common thread of this work. In addition, detailed Elliott analysis of temperature-dependent absorption spectra complements the investigations. The utilized Elliott formula describes the absorption spectrum of excitonic semiconductors and can be decomposed into a series of discrete excitonic resonances below the bandgap energy E_g arising from bound electron–hole pairs and an ionization continuum of unbound electrons and holes above E_g.[77–79] By least-squares fits of a generalized version of the Elliott formula to experimental spectra, additional characteristic parameters such as bandgap energy E_g and exciton binding energy E_B can be extracted.

The significant role of excitons with respect to integral properties and the working principle of perovskite solar cells is corroborated by results of combined temperature-dependent absorption and electroabsorption spectroscopy for the mostly studied model system methylammonium lead iodide, $MAPbI_3$. The good agreement of the energy of the fundamental optical resonance probed by electromodulation spectroscopy with the discrete excitonic contribution of the Elliott formula demonstrates its excitonic nature arising from bound electron–hole pairs instead of unbound continuum states. These findings hold true for temperatures from 10 K up to room temperature and were further substantiated by a detailed line-shape analysis of experimental electroabsorption spectra.

This rather proof-of-principle investigation was followed by a comparative study of different perovskite compositions utilizing the exciton binding energy as a quantitative measure. The experimental determination based on temperature-dependent absorption spectra was achieved by least-squares fitting of the Elliott formula and an additional alternative approach based on a so-called f-sum rule utilizing additional numerical integration.[352] The incorporation of bromine into $MAPbI_3$ allows tuning of the bandgap energy from 1.6 eV up to 2.3 eV but is accompanied by a significantly increased exciton binding energy which possibly hinders charge separation and collection in the device. However, for only small bromine contents, the partial replacement of the organic methylammonium cation with cesium and formamidinium – leading to so-called multiple-cation mixed-halide perovskites – results in values for the exciton binding energy even below the ones of standard $MAPbI_3$.

In general, excitonic effects are a double-edged sword with regard to photovoltaics. On the one hand, bound states of photogenerated electrons and holes are detrimental to an efficient separation and collection of these charge carriers and, thereby, reduce the solar cell's performance. However, on the other hand, a finite exciton binding energy significantly increases the absorption coefficient due to the so-called Sommerfeld enhancement[78, 348] which was illustrated on the basis of experimental absorption spectra of $MAPbI_3$ utilizing the Elliott formula. The obtained absorption and photocurrent spectra indicate that the exciton binding energy of perovskite solar cells is in the ideal range for both a significant Sommerfeld enhancement in absorption and still an efficient charge-carrier collection. This applies in particular to the exciton binding energies of $MAPbI_3$ and the optimized multiple-cation mixed-halide compounds in the order of thermal energies at room temperature which show excellent device performance.

Furthermore, absorption and electroreflectance spectroscopy were employed as facile tools for studies of temperature-induced phase transitions in different perovskite compounds due to the impact of structural changes on the bandgap energy. In particular, the suppression of phase transitions in multiple-cation compounds in the temperature range from 10 K to 300 K was detected and confirmed by X-ray diffraction measurements. These results might be interesting for low-temperature applications such as photovoltaics in space or foster further investigations of phase transitions in the standard solar-cell operation regime above room temperature.

Despite all outstanding characteristics of organic–inorganic perovskites that promise a bright future in photovoltaics, one key challenge remains which clearly is stability. One of the most prominent examples in this context is the halide segregation in mixed-halide compounds under illumination. Although additional incorporation of multiple cations enhances the intrinsic stability of mixed perovskites[143, 167, 336, 337] a reversible decrease of the bandgap energy under illumination and applied voltage was detected using electroreflectance spectroscopy as sensitive and reliable determination method. The exceptional reversibility suggests iodine–bromine segregation as origin of the observed bandgap instabilities which was further supported by the fact that bromine-free compositions proved to be stable. Furthermore, in-situ X-ray diffraction (XRD) and Time-of-Flight Secondary Ion Mass Spectrometry (ToF-SIMS) revealed reversible structural changes which were assigned to the formation of bromine-rich and iodine-rich domains under illumination and bias. An evaluation of possible driving forces identified that the bandgap decrease caused by the combined influence of illumination and bias clearly exceeds the individual contributions induced by light and voltage only. Thereby, the importance of additional electric fields in stability considerations is strongly highlighted. Moreover, the observed reversible bandgap instabilities cannot be avoided completely since illumination and bias are intrinsic to the solar cell's operation conditions. However, the extent of the bandgap decrease strongly increases under the combined influence of high relative humidity and high oxygen content which gives hope to reduce these detrimental effects by effective device encapsulation against oxygen and moisture.

Future Work and Outlook

Considering the potential of electromodulation spectroscopy as characterization tool for thin-film solar cells, a deeper understanding of the interplay of spectra detected at fundamental and second-harmonic frequency and electric fields in the device could open up the possibility of non-destructive investigations of built-in electric potentials in complete devices. The required in-depth computational studies of the influence of spatially varying electric fields in the device on electroreflectance spectra are currently under investigation using the simulation routines which have been newly developed or refined from previous studies.[74,75,397]

The role of excitons with regard to fundamental optical resonances probed by electromodulation spectroscopy has been elucidated within this work. However, further comprehensive examinations of their implications for solar-cell performance are needed, in particular, with respect to charge-carrier dynamics. Extended photocurrent spectroscopy studies can contribute to this research question, preferably combined with other characterization techniques. Spatially and temporally resolved photoluminescence spectroscopy allows the examination and quantification of charge-carrier transport.[205,482,483] Thereby, changes in transport properties of excitons compared to unbound electrons and holes could reveal additional insights and implications for photovoltaics.

The characterization of reversible bandgap instabilities in mixed perovskites caused by segregation into iodine-rich and bromine-rich domains raises various questions about the role of charge-extraction layers and interfaces, not least because of significant halide diffusion into spiro-OMeTAD detected by ToF-SIMS measurements. Consequently, solar cells with different hole-transport materials as well as layer thicknesses and doping levels should be thoroughly studied both experimentally and theoretically. Considering the experimental side, correlation with – preferably, spatially resolved – structural analysis is essential. Potentially, novel measurement protocols have to be established to guarantee fair quantitative comparison of different architectures. Moreover, the influence of relative humidity and oxygen merit a closer investigation, e.g., by correlation with compositional changes. Frequency-dependent measurements especially at very low frequencies could improve the understanding of the influence of the comparably slow ion migration and relevant time scales of these processes.

Furthermore, novel approaches to increase the long-term stability of perovskite solar cells, e.g., an additional layer of a two-dimensional perovskite[54,183–185] or completely inorganic devices,[484, 485] offer a broad field for further studies to gain deeper insights into perovskite instabilities and the consequences for device performance

Since long-term stability clearly is the key challenge of perovskite-based solar cells and related applications, a comprehensive understanding of mechanisms impairing their stability is crucial. By resolving still present instability and degradation issues, a bright future for organic–inorganic perovskites is guaranteed based on their outstanding optoelectronic properties and versatile fabrication methods.

Appendix A

Electro-Optic Characterization Techniques and Experimental Details

A.1 Verification of Temperature-Dependent Absorption Spectra by Means of a Commercial System with Integrating Sphere

In order to validate the results obtained by temperature-dependent absorption spectroscopy in this work, exemplary spectra obtained by the home-built setup used for temperature-dependent measurements and a commercial system with integrating sphere (*PerkinElmer Lambda 1050*) are compared. The latter have been measured together with Raphael Schmager at the *Light Technology Institute* (LTI) at *KIT*.

Figure A.1 (a) presents a direct comparison for a perovskite solar cell with $MAPbI_3$ absorber layer at room temperature. Both absorbance spectra A are calculated from transmittance measurements by $A = -\log(T)$. In the relevant photon-energy range (as depicted), nearly no deviations between the home-built setup used for temperature-dependent measurements (red squares) and the commercial system (black squares) are observable confirming the negligible influence of scattered light on the evaluation of the spectra.

In Fig. A.1 (b), transmittance T (red curve) and reflectance R (blue curve) are depicted for the same $MAPbI_3$-based solar cell measured by the commercial system mentioned above. Since R varies only moderately in the relevant energy range and is rather small compared to T, the approximation $A = -\log(T)$ for the absorbance is sufficiently justified – at least a high temperatures (as presented in chapter 4.1). However, at cryogenic temperatures, more pronounced features in reflectance could influence the evaluation.

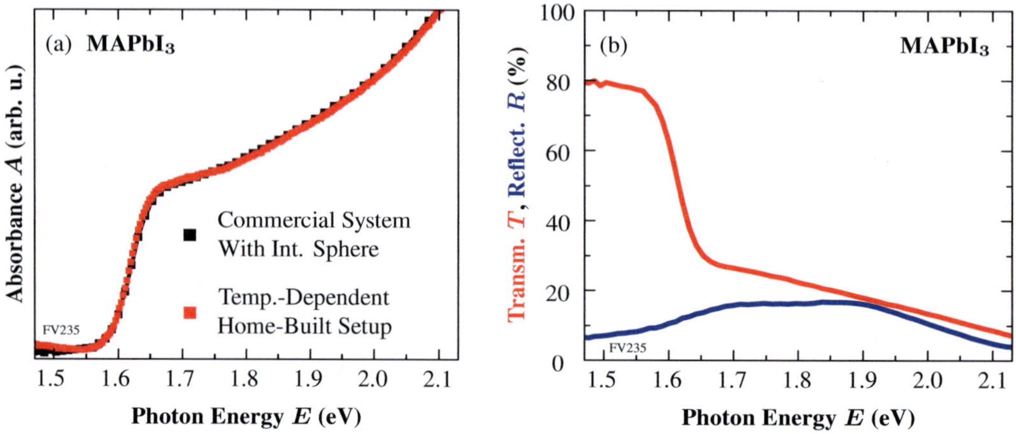

Figure A.1: (a) Exemplary comparison of absorbance $A = -\log(T)$ spectra measured by the temperature-dependent home-built setup used in this work (red squares) with results from a commercial *PerkinElmer Lambda 1050* system with integrating sphere (black squares) at room temperature. The excellent agreement of both spectra in the energy range depicted confirms the negligible influence of scattered light for the evaluation (Note that both spectra are scaled independently in arbitrary units since absolute values are neither relevant for this analysis nor precisely measurable with the home-built setup.). (b) Transmittance T (red curve) and reflectance R (blue curve) measured by the above-mentioned commercial setup. The only moderate variations and the comparably small values of R (with respect to T) sufficiently justify the evaluation of the absorbance $A = -\log(T)$ solely based on transmittance T. Data obtained together with Raphael Schmager at the *Light Technology Institute* (LTI) at *KIT*.

A.2 Measurement and Evaluation of Temperature-Dependent Photocurrent Spectroscopy

In this section, the basic idea including measurement principle and evaluation of temperature-dependent photocurrent spectroscopy discussed in chapter 7.3.2 is outlined. Details about the experimental setup and further results can be found in the master's thesis of Philip Lange.[422]

As presented in chapter 7.3.2, the spectrally-resolved *external quantum efficiency* (EQE) is defined as (see Eq. (7.3.2))

$$\mathrm{EQE}(\hbar\omega) = \frac{\text{\# of extracted charge carriers }(\hbar\omega)}{\text{\# of incident photons }(\hbar\omega)} = \frac{n_{\mathrm{e,h}}(\hbar\omega)}{n_{\mathrm{ph}}(\hbar\omega)}. \tag{A.1}$$

The number of extracted charge carriers $n_{\mathrm{e,h}}(\hbar\omega)$ during time Δt with spectrally-resolved photocurrent $I_{\mathrm{ph}}(\hbar\omega)$ and elementary charge e is given by

$$n_{\mathrm{e,h}}(\hbar\omega) = \frac{I_{\mathrm{ph}}(\hbar\omega)}{e}\Delta t. \tag{A.2}$$

Similarly, the number of incident photons $n_{\text{ph}}(\hbar\omega)$ in the interval Δt can be expressed as

$$n_{\text{ph}}(\hbar\omega) = \frac{E_{\text{in}}(\hbar\omega)}{\hbar\omega}\Delta t = \frac{P_{\text{in}}(\hbar\omega)}{A \cdot \hbar\omega}\Delta t \tag{A.3}$$

where $E_{\text{in}}(\hbar\omega) = P_{\text{in}}(\hbar\omega)/A$ and $P_{\text{in}}(\hbar\omega)$ denote the irradiance and power (impinging on area A) of the incident (monochromatic) light, respectively.

The experimental setup for the simultaneous measurements of temperature-dependent absorption and photocurrent spectra is, in principle, similar to the one used for temperature-dependent modulation spectroscopy described in chapter 6.1. In order to enhance the signal amplitude of the photocurrent measurement, a lock-in technique is applied. Therefore, a chopper wheel is introduced directly after the first monochromator (used as tunable light source) which provides the reference modulation signal for the lock-in amplifier. The solar-cell sample is mounted in the liquid-helium cryostat and connected via a transimpedance amplifier with the lock-in amplifier for sensitive photocurrent detection. By tuning the incident wavelength using the first monochromator, a photocurrent spectrum can be obtained. Simultaneously, the transmission spectrum of the (semitransparent) solar cell is recorded by the silicon detector (this time without the use of the second monochromator) to calculate the respective absorption spectrum.

The power spectrum $P_{\text{in}}(\hbar\omega)$ of the incident light is then determined by a silicon reference cell (*FDS100* by *Thorlabs*) which is mounted in the cryostat at the sample position before or after the actual measurement. The spectral responsitivity of the silicon cell given by the manufacturer is taken into account in the evaluation. However, since an accurate determination of the illuminated area A on the sample in the cryostat is not feasible, the calculation of precise values of the radiance $E_{\text{in}}(\hbar\omega)$ and, thereby, absolute EQE values is precluded.

In order to verify the reliability of the EQE results, measured spectra at room temperature are compared to similar ones obtained by a commercially available system (*Bentham PVE300*). An example of the results for the home-built setup used for temperature-dependent measurements (red) and a commercial EQE system (black) is depicted in Fig. A.2 (It should be noted that the red curve is scaled arbitrarily in vertical direction due to the unfeasible quantitative calibration of the home-built setup.). The comparison indicates a shift in photon energy of about 15 meV which could be partly related to different ambient temperatures, but, more probably, is caused by not ideal wavelength calibration of the monochromator.[422] However, this does not influence the conclusions in chapter 7.3.2 since only simultaneously measured spectra are compared using the same monochromator for wavelength tuning. Despite this small shift, both spectra match very well which can be easily seen by shifting the spectrum of the home-built setup by 15 meV (dashed red curve). Therefore, the reliability of the home-built setup is confirmed.

Figure A.3 presents temperature-dependent EQE spectra of $MAPbI_3$ showing the strong increase of photocurrent and EQE of more than three orders of magnitude approaching room temperature (compared to $T = 10\,\text{K}$) which rationalizes the efficient charge-carrier collection under solar-cell working conditions despite the observable excitonic signatures at cryogenic temperatures (see chapter 7.3.2).

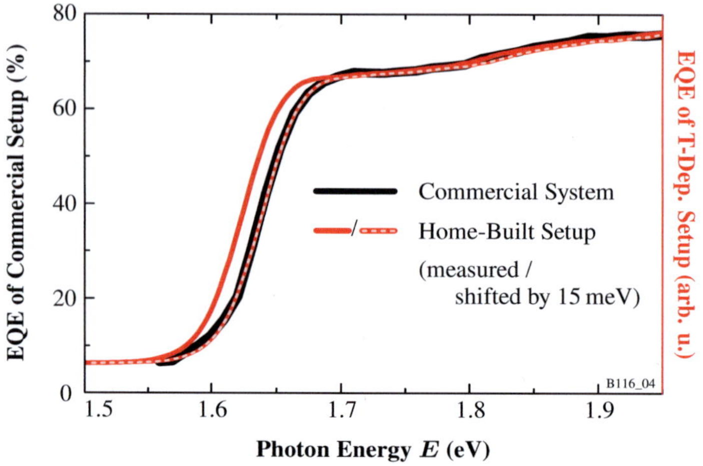

Figure A.2: Comparison of EQE spectra at room temperature measured by the home-built setup (red curve) used for low-temperature EQE measurements in chapter 7.3.2 and a commercially available system (Bentham PVE300, black curve). The red curve is scaled arbitrarily in vertical direction due to the unfeasible quantitative calibration of the home-built setup which would require a precise determination of the illuminated area A on the sample inside the cryostat. The shift in photon energy of about 15 meV is probably caused by not ideal wavelength calibration of the monochromator. However, this is not relevant for the relative comparison of simultaneously obtained absorption and EQE spectra in chapter 7.3.2. Despite from this shift, both spectra match very well verifying the reliability of the home-built setup. Data published in [422].

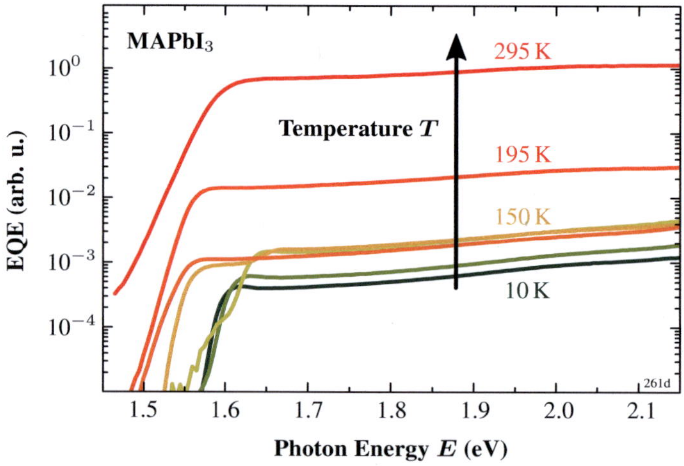

Figure A.3: EQE spectra of $MAPbI_3$ for temperatures $T = 10\,K - 295\,K$. The photocurrent and EQE increase more than 3 orders of magnitude for increasing temperature from 10 K up to room temperature rationalizing the good photovoltaic performance of corresponding solar cells. Data published in [422].

A.3 Optical and Electronic Components of the Experimental Setups

The following tables list all optical and electronic components of the used experimental setups for absorption (see chapter 4), electromodulation (see chapter 6), and photocurrent (see chapter 7.3.2 and appendix A.2) spectroscopy.

Absorption Spectroscopy

A schematic overview of the experimental setup used for the measurement of temperature-dependent absorption spectra of perovskite thin films is depicted in Fig. 4.1.

Component	Focal Length [mm]	Diameter [mm]	Comment
L1	80	38	Quartz
L2	80	30	Quartz
L3	100	30	Achromatic
L4	150	50	Achromatic
L5	100	38	Quartz

Table A.1: Optical components used for temperature-dependent absorption spectroscopy of perovskite thin films as depicted in Fig. 4.1.

Part	Manufacturer	Type	Comment
Halogen Lamp			$P = 150\,\mathrm{W}$
Cryostat	Cryo Industries	Model 600	Liquid-helium cryostat
Monochromator	Acton Research Corp.	SpectraPro-275	Grating monochr. with $f = 0.275\,\mathrm{m}$
CCD Camera	IKS Optoelektronics	XMCD	

Table A.2: Electronic components and additional equipment used for temperature-dependent absorption spectroscopy of perovskite thin films as presented in Fig. 4.1.

Electromodulation and Photocurrent Spectroscopy

Figures 6.1 and 6.2 present a schematic overview of the experimental setups for electromodulation spectroscopy at room temperature and cryogenic temperatures. Additional modifications for the measurement of temperature-dependent photocurrent spectra are discussed in appendix A.2.

The following tables are similarly structured in three sections summarizing components used for room-temperature electromodulation spectroscopy, additional equipment for low-temperature measurements, and further components used for photocurrent spectroscopy.

Component	Focal Length [mm]	Diameter [mm]	Comment
F1	–	–	OG515, GG435
L1	300	88.9	Glass
L2	300	31.5	Achromatic
L3	100	50.8	Glass
L4	60	50.8	Glass
L5	150	25.4	Quartz
L6	75	50.8	Glass
L7	100	50.8	Achromatic
L8	80	50.8	Achromatic
L9	100	30.0	Achromatic
L10	150	50.8	Achromatic
L6* (replaces L6)	120	38.1	Achromatic

Table A.3: Optical components used for electromodulation spectroscopy at room temperature (see Fig. 6.1) and cryogenic temperatures (see Fig. 6.2) as well as photocurrent spectroscopy (see chapter 7.3.2 and appendix A.2).

Part	Manufacturer	Type	Comment
Halogen Lamp			$P = 250\,\mathrm{W}$
TRIAX Monochromator	Jobin Yvon	TRIAX 320	Used as tunable light source 600 lines/mm, $f = 0.32\,\mathrm{m}$ typ. slit width $(0.7 - 1.0)\,\mathrm{mm}$
Si Photodiode	Electro-Optical Systems Inc.	UVS-025-TE2-H	TE-cooled (PS / TC-1)
Lock-In Amplifier (×2)	Princeton Applied Research	EG & G 5209	Used for "ΔR"/"ΔT" detection
Digital Multimeter	Keithley	Multimeter 2000	Used for "R"/"T" detection
Function Generator	Tektronix	AFG 3102	Generation of modulation voltage
Hygrometer	Extech Instruments	Model 445815	
Solar Simulator	LOT Quantum Design		Low-cost solar simulator
Liquid-Helium Cryostat	Cryo Industries	80N1044	Variable-temperature cryostat
Monochromator HR640	Jobin Yvon	HR640	Used for suppression of stray light and luminescence, $f = 0.64\,\mathrm{m}$
Chopper	Stanford Research Systems Inc.	SR-540	Optical Chopper
Transimpedance Amplifier	EG & G Optoelectronics		Used to convert photocurrent to voltage

Table A.4: Electronic components and additional equipment used for electromodulation spectroscopy at room temperature (see Fig. 6.1) and low temperatures (see Fig. 6.2) as well as photocurrent spectroscopy (see chapter 7.3.2 and appendix A.2).

Appendix B

Reliability of Elliott Fits and f-Sum Evaluation Method

The evaluation of experimental absorption spectra using the generalized Elliot formula introduced in chapter 4 can be expected to yield reliable results, in particular, at cryogenic temperatures and/or high exciton binding energies since in these cases the pronounced excitonic peak in the spectrum facilitates precise fitting (see, e.g., Fig. 7.1). However, for materials with small exciton binding energy E_B such as Cs0.05 ($Cs_{0.05}(FA_{0.83}MA_{0.17})_{0.95}Pb(I_{0.83}Br_{0.17})_3$ as defined in chapter 7.2.3) and large broadening Γ at room temperature, the excitonic peak becomes indistinct which results in a compromised reliability of the fit. This is demonstrated in Fig. B.1 by three different fitting solutions with reasonable agreement with the experimental data but clearly varying E_B values. The range of 12 meV up to 24 meV provides a rough measure for the fit uncertainty under these conditions.

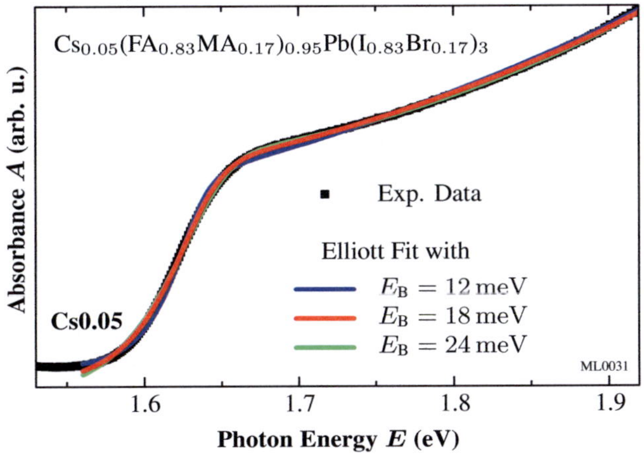

Figure B.1: Generalized Elliott fits to a room-temperature absorption spectrum of Cs0.05. Because of the large broadening Γ at room temperature and, at the same time, small exciton binding energy E_B, no clear excitonic peak is visible. As a result, the reliability of the Elliott fit is compromised which is illustrated by three different fitting solutions. All three curves apparently provide a decent agreement with the experimental spectrum, but correspond to fairly different values for E_B ranging from 12 meV up to double the value.

Figure B.2 displays the temperature dependence of the non-parabolicity $\left(8\mu^2 b\right)/\hbar^4$ for the perovskite compounds investigated in chapter 7.2.3 which are extracted from Elliott fits to absorption spectra. In particular, for both Cs-compounds Cs0.05 (purple dots) and Cs0.10 (blue dots, $Cs_{0.1}FA_{0.765}MA_{0.135}Pb(I_{0.765}Br_{0.235})_3$ as defined in chapter 7.2.3), the values for the non-parabolicity factor $\left(8\mu^2 b\right)/\hbar^4$ do not show significant variations with temperature. In the high-temperature range ($T > 150\,\mathrm{K}$) used for the f-sum evaluation (dashed boxes), the deviations are below $\pm 3\%$ and $\pm 2\%$ of the average value for Cs0.05 and Cs0.10, respectively. While slightly increased to $\pm 5\%$ and $\pm 10\%$ in case of $MAPbI_3$ and $MAPbBr_3$, respectively, the variations are still reasonably low providing a reliable f-sum evaluation.

The uncertainty of the exciton binding energy E_B solely due to variations of the non-parabolicity factor are below $\pm 2\,\mathrm{meV}$ (even below $\pm 1\,\mathrm{meV}$ for both Cs-compounds). In addition, deviations of the integrated normalized absorption I_{exp} increase the uncertainty to about $\pm 3\,\mathrm{meV}$ (up to $\pm 5\,\mathrm{meV}$ for $MAPbBr_3$) which is still notably lower than the range of $12\,\mathrm{meV} - 24\,\mathrm{meV}$ shown in Fig. B.1. Certainly, this discussion is not meant to be a complete quantitative assessment since many uncertainties of the measurement technique affecting both evaluation methods are not included. However, it emphasizes the improved reliability of the f-sum rule method compared to individual Elliott fits for determination of E_B close to room temperature.

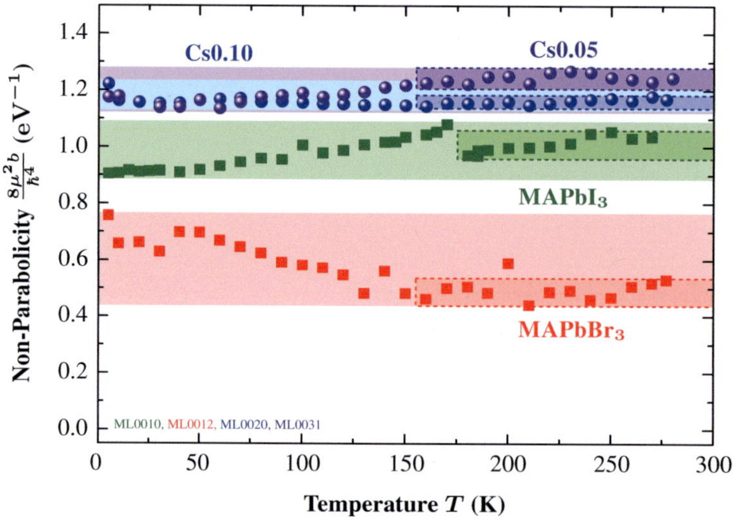

Figure B.2: Temperature dependence of the non-parabolicity factor $\left(8\mu^2 b\right)/\hbar^4$ for Cs0.05 (purple dots), Cs0.10 (blue dots), $MAPbI_3$ (green squares), and $MAPbBr_3$ (red squares) determined by fitting of Elliott fits to absorption spectra. The overall variations are rather small, especially for both Cs-compounds, providing reliable results of the f-sum evaluation. The dashed boxes depict the temperature range $T > 150\,\mathrm{K}$ used for f-sum determination of high-temperature E_B values discussed in chapter 7.2.3. They represent variations of $\pm 3\%$ (Cs0.05), $\pm 2\%$ (Cs0.10), $\pm 5\%$ ($MAPbI_3$), and $\pm 10\%$ ($MAPbBr_3$) of the respective average value in this temperature range emphasizing the suitability for reliable results of the f-sum rule method.

Appendix C

Analysis of Electromodulation Spectra Using First-Derivative Functional Form Line-Shape Fits

In contrast to the rather simple and explicit form of the Third-Derivative Functional Form (TDFF) describing band-to-band transitions of (quasi-)free charge-carriers (see chapter 5.3.1), more care must be taken in case of the description of discrete resonances using a First-Derivative Functional Form (FDFF). An individual resonance is characterized by its energy E, line-width broadening Γ and oscillator strength f (or intensity I) which in principle can all be modulated by an electric field and result in respective contributions $f_E^{(\xi)}$, $f_\Gamma^{(\xi)}$, $f_I^{(\xi)}$ of the FDFF line shape (see chapter 5.3.2). In addition, line-width broadening can be caused by different mechanisms and is often modeled by either Lorentzian or Gaussian broadening. The respective mathematical expressions are given in the following section which is followed by further discussions of relevant modulation mechanisms and respective FDFF line shapes for curve fitting in section C.2 (in addition to chapter 7.1.2).

C.1 Mathematical Expressions of Lorentzian- or Gaussian-Broadened FDFF Line Shapes

The modulation terms $f_E^{(\xi)}$, $f_\Gamma^{(\xi)}$, $f_I^{(\xi)}$ of the First-Derivative Functional Form and dielectric functions ϵ assumed for Lorentzian or Gaussian broadening are listed on the next page together with their visualization in Fig. C.1 (identical with Fig. 5.5).[356, 381]

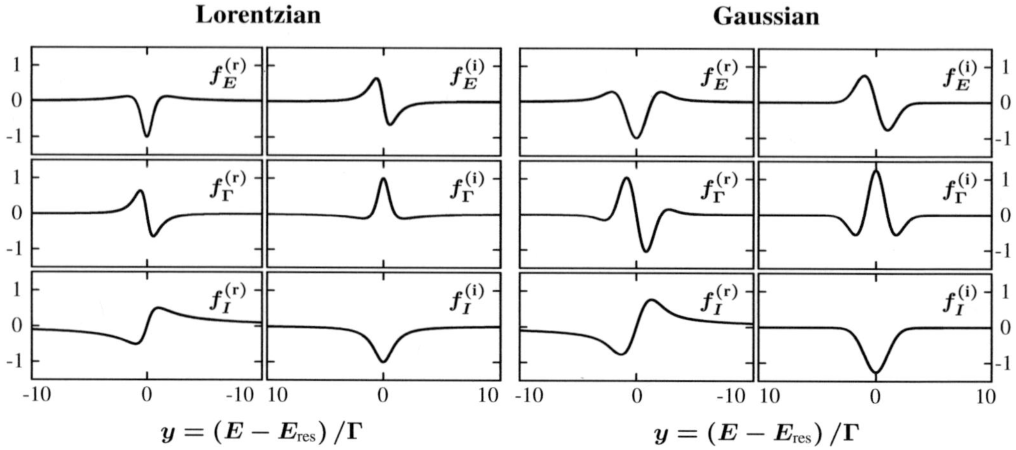

Figure C.1: Calculated line shapes for the derivatives of the real and imaginary part of the dielectric function for (left) Lorentzian and (right) Gaussian broadening with respect to resonance energy E_{res}, broadening Γ and intensity I. Identical with Fig. 5.5.

Lorentzian

$$\epsilon = 1 + \frac{I}{E - E_{res} + i\Gamma}$$

$$\begin{aligned}
f_E^{(r)} &= \left(y^2 - 1\right) / \left(y^2 + 1\right)^2 \\
f_E^{(i)} &= -2y / \left(y^2 + 1\right)^2 \\
f_\Gamma^{(r)} &= f_E^{(i)} \\
f_\Gamma^{(i)} &= -f_E^{(r)} \\
f_I^{(r)} &= y / \left(y^2 + 1\right) \\
f_I^{(i)} &= -1 / \left(y^2 + 1\right)
\end{aligned} \tag{C.1}$$

Gaussian

$$\epsilon = 1 + I\,(L_1 + iL_2)$$

$$\begin{aligned}
f_E^{(r)} &= -\Phi\left(1, 1/2, -y^2/2\right) \\
f_E^{(i)} &= -\left(\pi/2\right)^{1/2} y \exp\left(-y^2/2\right) \\
f_\Gamma^{(r)} &= -2y\,\Phi\left(2, 3/2, -y^2/2\right) \\
f_\Gamma^{(i)} &= -\left(\pi/2\right)^{1/2} \left(y^2 - 1\right) \exp\left(-y^2/2\right) \\
f_I^{(r)} &= +y\,\Phi\left(1, 3/2, -y^2/2\right) \\
f_I^{(i)} &= -\left(\pi/2\right)^{1/2} \exp\left(-y^2/2\right)
\end{aligned} \tag{C.2}$$

with

$$\begin{aligned}
y &= \left(E - E_{res}\right)/\Gamma, \\
L_1 &= +\left(y/\Gamma\right)\,\Phi\left(1, 3/2, -y^2/2\right), \\
L_2 &= -\left(\pi/2\right)^{1/2} \left(1/\Gamma\right) \exp\left(-y^2/2\right),
\end{aligned} \tag{C.3}$$

where $\Phi(a, b, x)$ denotes the confluent hypergeometric function.

C.2 Comparison of Different Contributions in FDFF Line-Shape Fits of Experimental Electromodulation Spectra

The different FDFF modulation terms f_E, f_Γ, f_I are fitted separately to measured electromodulation spectra in order to assess how well each of them reproduces the experimental data. Thereby, the influence of the individual contributions on the overall quality of the fit can be investigated. Figure C.2 serves as an example of a $\Delta T/T$ electroabsorption spectrum measured on a semitransparent solar cell with $MAPbI_3$ absorber layer at temperature $T = 260\,\text{K}$.

Independent of the type of broadening, the intensity-modulation term f_I is clearly not an appropriate line shape since it provides by far the worst agreement with experimental data. Therefore, f_I is neglected in the line-shape fits presented within this work.

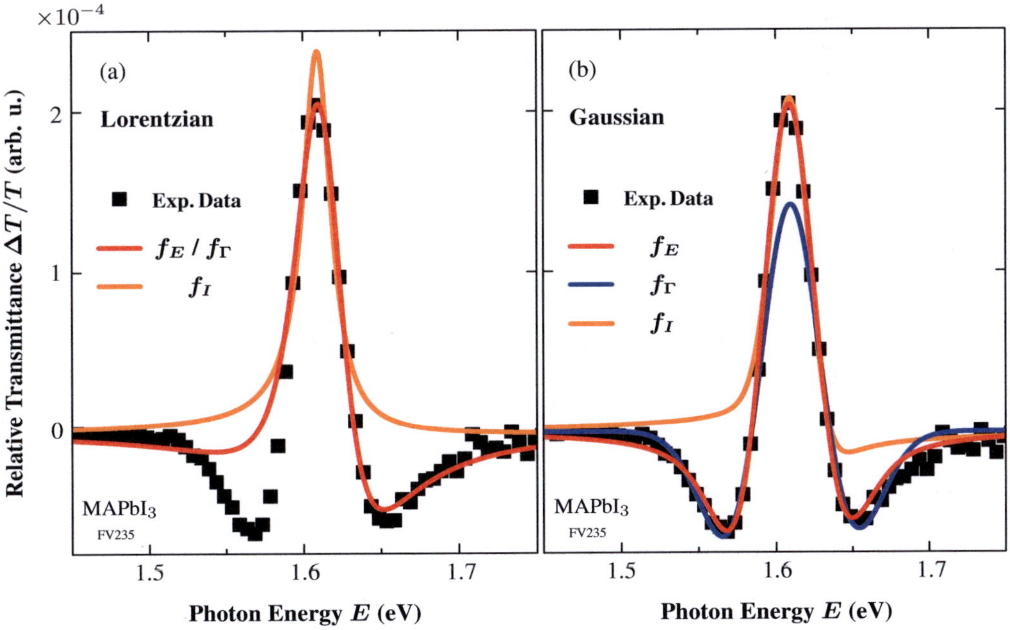

Figure C.2: Least-squares fits of individual FDFF modulation terms f_E, f_Γ, f_I including (a) Lorentzian or (b) Gaussian broadening on a measured $\Delta T/T$ electroabsorption spectrum of a semitransparent solar cell with $MAPbI_3$ absorber layer at temperature $T = 260\,\text{K}$. For both broadening types, the intensity-modulation terms f_I are apparently not well-suited. Due to the equivalent form of f_E and f_Γ (see chapter 5.3.2 and equation (C.1)) for Lorentzian broadening, effectively only one line shape remains for fitting. In case of Gaussian broadening, both f_E and f_Γ can be utilized. Original measured spectra obtained in collaboration with Alice Magin.[389]

As a result, effectively only the energy-modulation term f_E remains as independent line shape for curve fitting in case of Lorentzian broadening (see Fig. C.2 (a)) since the form of f_E and f_Γ are equivalent (see chapter 5.3.2 and equation (C.1)). In addition, the residual line shape f_E can be expressed by (5.21) with $n = 2$ as presented in chapter 5.3.2.[356]

In case of Gaussian-broadened FDFF line shapes, f_E and f_Γ represent independent modulation terms and both can be utilized in curve fitting as can be seen in Fig. C.2 (b) (Still, approximated expressions are possible as discussed in chapter 5.3.2.). Here, linear combinations of both f_E or f_Γ do not show significantly improved fit results and are therefore not further pursued. Instead, line shape fits of f_E and f_Γ to $\Delta T/T$ of a $MAPbI_3$-based solar cell are compared for different temperatures. Figure C.3 (a) presents the *coefficient of determination* R^2 representing the quality of the fits in both cases. While the differences are small representing good agreement with measured data for both line shapes, a striking trend of improved fit quality toward f_E contributions at high temperatures and f_Γ modulation terms at low temperatures is observable. Exemplary spectra and fits at $T = 130\,\mathrm{K}$ and $T = 260\,\mathrm{K}$ in Fig. C.3 (b) and (c), respectively, also show this behavior.

In principle, these observations could represent a general trend favoring broadening-modulation terms at lower temperatures and energy-modulation at room temperature in case of Gaussian-broadened FDFF line shapes. However, in case of samples with $Cs_xFA_yMA_{1-x-y}Pb(I_{1-z}Br_z)_3$ absorber layer, no overall trend is observable. Therefore, the excellent agreement of the crossover from f_E to f_Γ line shapes with the phase-transition temperature of 162 K in $MAPbI_3$ indicates that these changes are presumably related to different crystal structures with different symmetries. Since changes in the crystal structure influence the wavefunction overlap especially of the inorganic I–Pb framework they are able to modify the electronic structure and, in turn, the response on a modulating electric field. Thereby, different modulation mechanisms can become dominant for different crystal structures, here modulation of line-width broadening and resonance energy in the low-temperature orthorhomic phase and room-temperature tetragonal crystal phase, respectively. Accordingly, the absence of a similar trend in $Cs_xFA_yMA_{1-x-y}Pb(I_{1-z}Br_z)_3$ perovskites matches the phase-transition suppression in these compounds demonstrated in chapter 8. Still, further detailed investigations are needed in order to verify and explain the observed trends.

In summary, based on FDFF line-shape fits, intensity modulation (described by f_I) is clearly not a relevant mechanism in case of the investigated perovskites and, therefore, neglected within this work. Since the remaining mathematical expressions f_E and f_Γ representing modulation of resonance energy and line-width broadening, respectively, are equivalent in case of Lorentzian broadening one of them is sufficient for FDFF curve fitting (and it is not possible to discriminate between both mechanisms). In case of Gaussian-broadened FDFF line shapes, the modulation of the resonance energy is often assumed to be the dominant mechanism (as discussed in the framework of *Stark theory* in chapter 5.2.3) which can be confirmed, e.g., for $MAPbI_3$ due to better agreement of the f_E modulation term (compared to f_Γ) with experimental spectra – at least for temperatures from $T = 160\,\mathrm{K}$ up to room temperature. However, below the phase-transition temperature $T_\mathrm{p} \approx 160\,\mathrm{K}$ from the tetragonal to the low-temperature orthorhombic phase, broadening-modulation seems to become the dominant mechanism which can be rationalized by changes in the electronic structure due to different crystal phases. In any case, for Gaussian-broadened FDFF curve fits, the modulation term (f_E or f_Γ) which reproduces the experimental spectra better is used for the evaluation within this work. A detailed comparative examination of Lorentzian- and Gaussian-broadened FDFF as well as TDFF line-shape fits is presented in chapter 7.1.2.

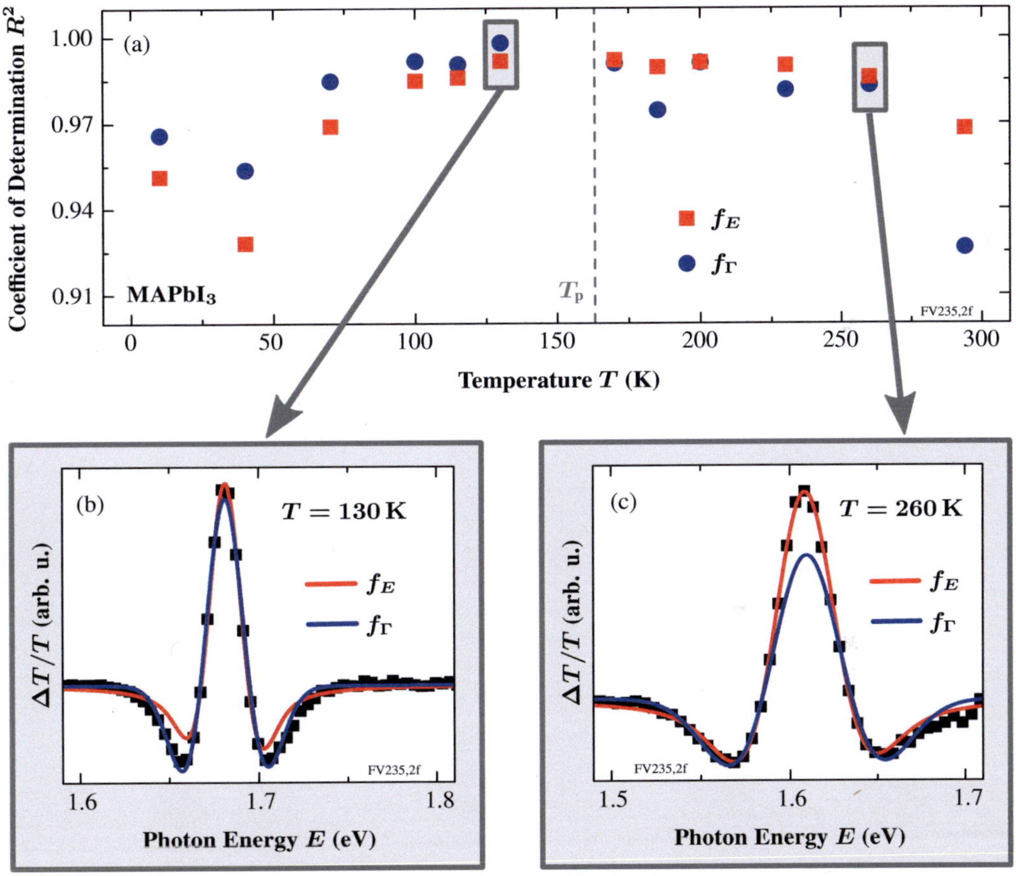

Figure C.3: (a) Coefficient of determination R^2 for Gaussian-broadened FDFF line-shape fits of temperature-dependent electromodulation spectra of $MAPbI_3$-based solar cells using either energy- or broadening-modulation terms, f_E (red squares) or f_Γ (blue circles), respectively. Despite the small differences, there is a striking trend of better agreement towards f_E contributions at high temperatures and f_Γ modulation terms at low temperatures. These observations are also reflected by exemplary spectra at (b) $T = 130\,\mathrm{K}$ and (c) $T = 260\,\mathrm{K}$. Original measured spectra obtained in collaboration with Alice Magin.[389]

Appendix D

Numerical Simulation of Electroreflectance Spectra

The simulation routine for the numerical calculation of electroreflectance spectra at fundamental and second-harmonic frequency detection is illustrated in Fig. D.1. Further information about the general simulation procedure can be found in [389].

The "electronic" part of the calculations is performed using the one-dimensional simulation tool *SCAPS* developed at the *University of Gent*.[394–396] These simulations require the definition of all layer thicknesses and electronic properties of the layer stack including, e.g., bandgap energies, doping concentrations, and trap densities. The calculation of j–V characteristics (using *SCAPS*) and comparison with experimental data can be used as a cross-check of the parameter set. The modulation voltage $V(t)$ is discretized into $V(t_n)$ with evenly-spaced times t_n and, for each voltage value $V(t_n)$, the respective electric field distribution in the device is computed. An example for a Cu(In,Ga)Se_2-based solar cell is depicted in Fig. D.2. In the space–charge region in the Cu(In,Ga)Se_2 absorber layer (starting at the interface to the CdS buffer layer), the field clearly shows a linear behavior which could be used for simplification of the simulation routine. As expected, an additional reverse bias of $V_{dc} = -2\,\mathrm{V}$ leads to an increase of the electric field in the device.

From the electric field distribution, the electro-optic energy $\hbar\theta$ can be calculated for each spatial position which then enters the optical properties of the respective material via the electro-optic function (see chapter 5.2). Clearly, the most relevant layer in this context is the absorber (and eventually the CdS buffer). For all other layers, the influence of the electric field can be neglected (also reflected by the electric field profile in Fig. D.2). In the simulation algorithm, the space–charge region is discretized in N layers and a constant electro-optic energy $\hbar\theta$ is assumed for each layer (see Fig. D.3).

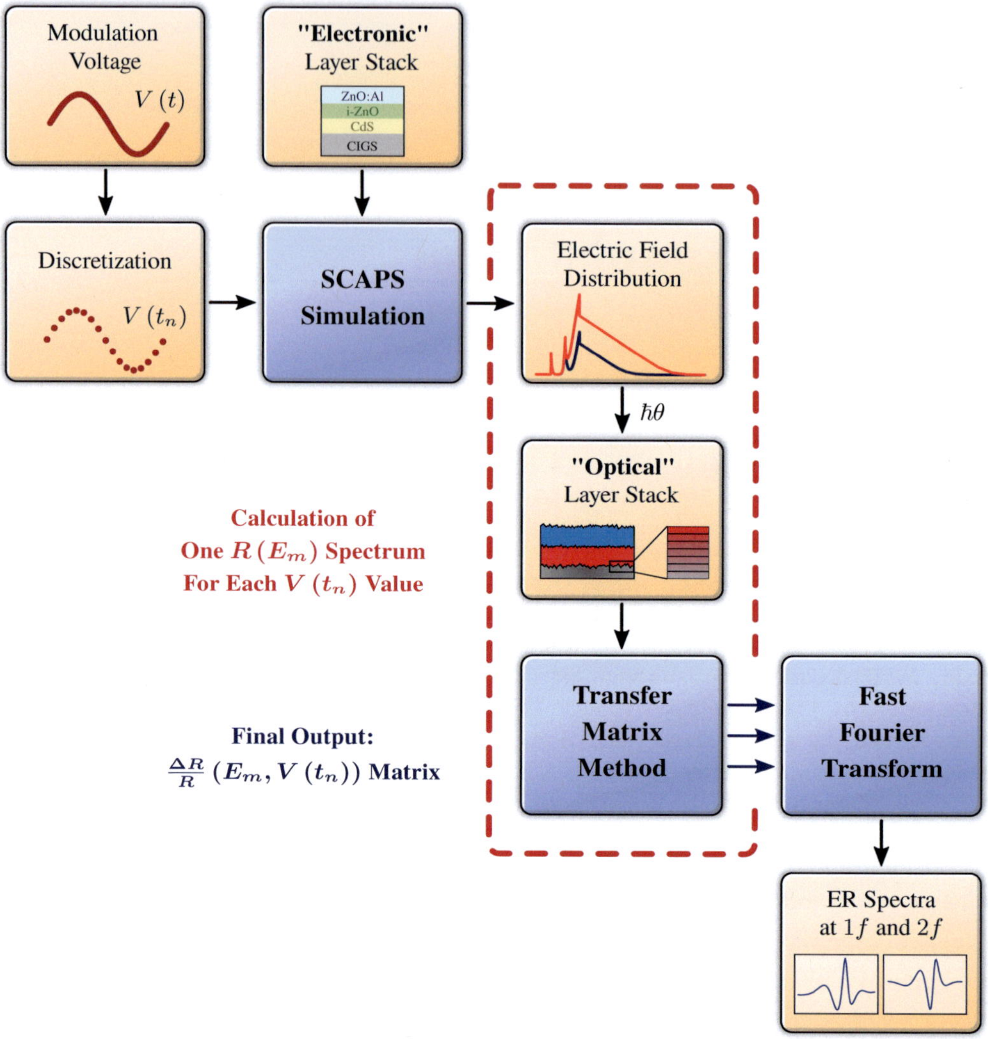

Figure D.1: Schematic Illustration of the simulation routine to calculate electroreflectance spectra at $1f$ and $2f$ detection. The simulation tool *SCAPS* is used to calculate one-dimensional electric field distributions in the device for different values $V(t_n)$ of the discretized modulation voltage $V(t)$. The resulting electric fields enter the optical parameters (n, κ) of the materials (here, only of the absorber layer) via the electro-optic energy $\hbar\theta$. Based on this "optical" layer stack, a reflectance spectrum $R(E_m)$ with discrete photon energies E_m is calculated for each $V(t_n)$ value using a transfer-matrix method algorithm. The overall output matrix $\frac{\Delta R}{R}(E_m, V(t_n))$ including all photon energies E_m of the spectra and the full discretized modulation waveform $V(t_n)$ is transformed using a *Fast-Fourier Transform* (FFT) to obtain electroreflectance spectra at frequencies $1f$ and $2f$. See also [389].

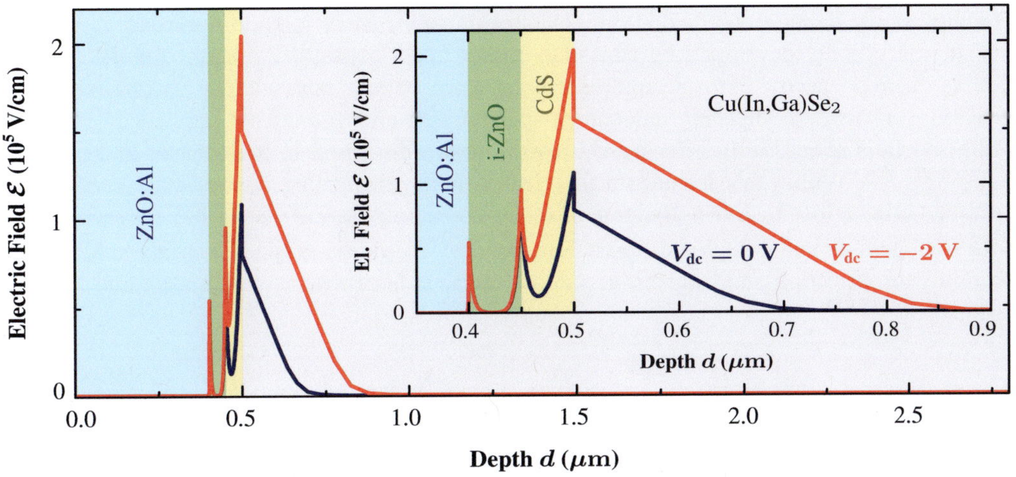

Figure D.2: Examplary simulated electric field distribution (and enlargement in the inset) of a thin-film solar cell with ZnO:Al / i-ZnO window layer, CdS buffer, and $Cu(In,Ga)Se_2$ absorber layer using SCAPS. A reverse bias of $V_{dc} = -2$ V (red curve) increases the electric field strength in the absorber layer. Data obtained in collaboration with Ellen Förstner.[486]

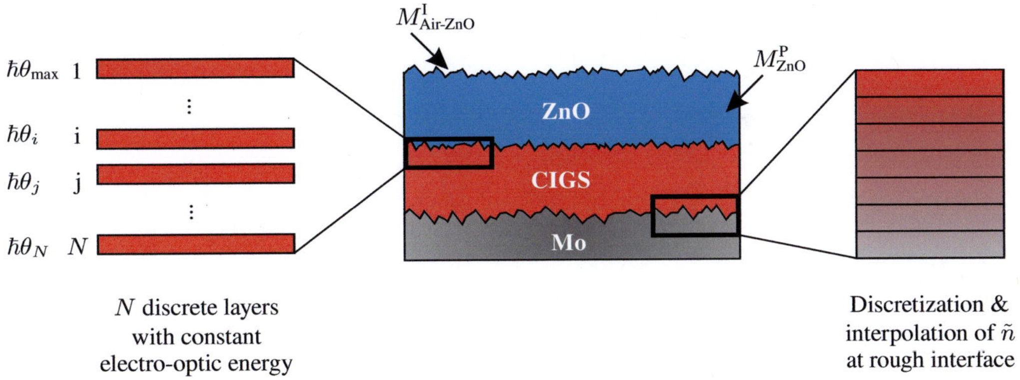

Figure D.3: Schematic illustration of the transfer-matrix method used for the calculation of reflectance spectra. The one-dimensional approach uses 2×2 interface and propagation matrices M^{I} and M^{P}, respectively. To implement the influence of an electric field, the stack is divided into N discrete layers with constant electro-optic energy $\hbar\theta$. Interface roughness is modeled by interpolating the refractive index $\tilde{n}$ over a certain distance. From [398].

For the optical simulations, a one-dimensional transfer-matrix method approach is utilized which was initially developed by Christian Huber and Christoph Krämmer for the simulation of $\Delta R/R$ spectra with an applied square-wave voltage and without second-harmonic detection (basically comparing reflectance spectra with and without an applied dc voltage). Further details about the transfer-matrix algorithm are discussed in their theses and publication.[74, 397, 398] Within this formalism the electric field propagating in the device (with wave vector k) is described by a two-component vector and a propagation matrix M_j^{P} for each layer j with thickness d_j and complex refractive index $\tilde{n}_j$.[398] Additionally an interface matrix M_{ij}^{I} accounts for reflection and transmission at an interface between two materials i and j in the layer stack:[398, 487]

$$M_j^{\mathrm{P}} = \begin{pmatrix} \mathrm{e}^{-\mathrm{i}k\tilde{n}_j d_j} & 1 \\ 1 & \mathrm{e}^{\mathrm{i}k\tilde{n}_j d_j} \end{pmatrix}, \quad M_{ij}^{\mathrm{I}} = \frac{1}{t_{ij}} \begin{pmatrix} 1 & r_{ij} \\ r_{ij} & 1 \end{pmatrix} \tag{D.1}$$

with the Fresnel coefficients t_{ij} and r_{ij} for transmission and reflection, respectively. Interface roughness between two layers is implemented by interpolation of the refractive indices of both materials for a certain thickness (see Fig. D.3). The propagation of light through the total layer stack is determined by multiplication of the individual matrices for each interface and material. Subsequently, the reflectance and transmittance of the complete stack can be calculated.

The respective reflectance spectrum $R\,(E_m)$ with discrete photon energies E_m is computed for each voltage value $V\,(t_n)$ (based on the corresponding electric field distribution and, thereby, "optical" layer stack) using the transfer-matrix algorithm. After calculating the relative change $\Delta R/R$ (compared to the reference reflectance spectrum at the average voltage V_{dc}), this results in a total matrix $\frac{\Delta R}{R}\,(E_m, V\,(t_n))$ including the spectra for all photon energies E_m and all voltage values $V\,(t_n)$ of the discretized modulation waveform.

Finally, a line-by-line *Fast Fourier transform* (FFT) is applied to the total matrix $\frac{\Delta R}{R}\,(E_m, V\,(t_n))$. Thereby, the time-domain spectrum of $\Delta R/R\,(t_n)$ (via $V\,(t_n)$) for each photon energy E_m is converted to a frequency distribution $\Delta R/R\,(f_n)$. By read-out of the respective columns representing frequencies f and $2f$ of the total output matrix of the line-by-line FFT, the desired electroreflectance spectra for $1f$ and $2f$ detection are obtained (illustrated in Fig. D.4). See [389] for further details on this as well as a discussion of suitable simulation parameters.

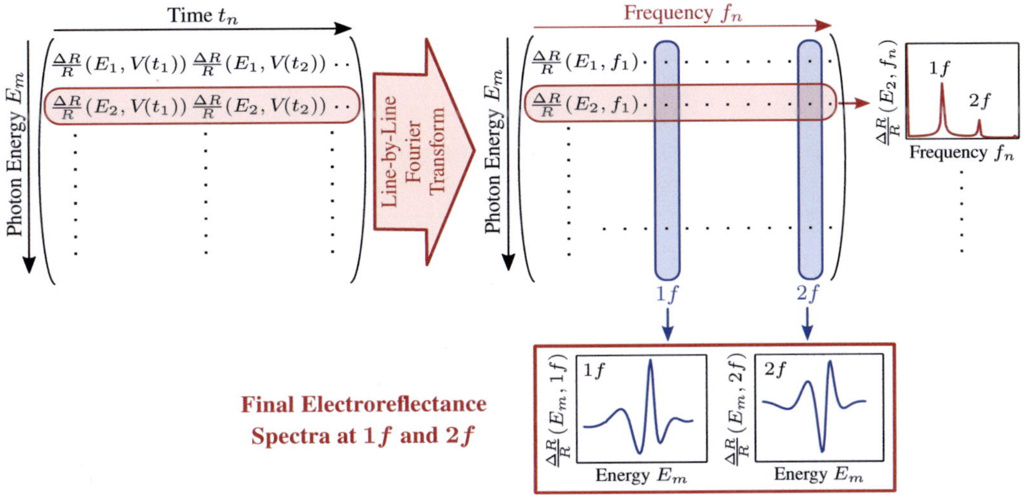

Figure D.4: Visualization of the applied line-by-line Fourier-transform method. The input matrix $\frac{\Delta R}{R}(E_m, V(t_n))$ for all discretized photon energies E_m of the spectra and all voltage values $V(t_n)$ of the discretized waveform is calculated by the transfer-matrix algorithm. Each line representing a certain photon energy E_m is Fourier transformed yielding a frequency distribution of $\frac{\Delta R}{R}(f_n)$. The respective columns of the output matrix $\frac{\Delta R}{R}(E_m, f_n)$ representing frequency $1f$ and $2f$ contain the desired electroreflectance spectra at fundamental and second-harmonic frequency. Adapted from [389].

Appendix E

Structural Characterization Methods and Experimental Details

E.1 X-Ray Diffraction Analysis and Experimental Instrumentation

X-ray diffraction analysis is based on coherent elastic scattering of X-rays at electrons in a material. X-rays scattered at different atomic planes in a crystal interfere constructively for distinct incident angles θ. By measuring the angle 2θ between incident and diffracted direction, the spacing d of two atomic planes can be deduced from Bragg's law:[353, 488]

$$2d \sin\theta = n\lambda \tag{E.1}$$

with wavelength λ of the incident X-rays and a positive integer n. This rather simple formula provides the basis for the interpretation of X-ray diffractograms which are mainly angular scans of scattered X-ray intensity over diffraction angle 2θ (with fixed θ–2θ relation of incident and scattered direction, so-called *theta-2theta scans*).[353] The specific set of distinct reflexes in experimental diffraction patterns and the corresponding d values represent "unique fingerprints" which enable the determination of the crystal structure and chemical composition of crystalline materials.[353] In practice, measured diffractograms are compared with theoretically predicted diffraction patterns of different materials. See, e.g., [353] for further details.

Experimental Instrumentation Used in This Work

In-situ X-ray diffractograms were obtained at the *Ludwig-Maximilian University* (LMU) in Munich by a *Bruker D8 Discover* X-ray diffractometer operating at 40 kV and 30 mA, employing Ni-filtered Cu-$K_{\alpha 1}$ radiation ($\lambda = 1.5406$ Å) and a position-sensitive *LynxEye* detector under ambient conditions in air (with $(34 \pm 5)\%$ relative humidity at a temperature of (24 ± 2) °C). Spectra are smoothed by a Savitzky-Golay algorithm. No shifts of the XRD reflexes to lower angles due to local heating resulting from the 1 sun illumination were found (which can be estimated to be in the order of $0.001\,°/\,\mathrm{K}$ and $0.002\,°/\,\mathrm{K}$ for the $14.1°$ and $28.4°$ perovskite peak, respectively) excluding distortions of the results due to thermal effects. See also [241].

The temperature-dependent X-ray powder diffraction measurements were performed by Bettina Rendenbach at the *Ludwig-Maximilian University of Munich* (LMU) on a *Huber G670 Guinier Imaging Plate* diffractometer with Co–$K_{\alpha 1}$ radiation ($\lambda = 1.788\,965$ Å) and a Ge(111) primary monochromator. Temperature control is realized by a low-temperature device 670.4, a closed-cycle He cryostat (*CTICryogenics*, model 22), and a temperature controller (*Lakeshore*, model 331). The perovskite powders were obtained by scraping off thin-films from their glass substrate and were prepared with grease between two foils. For direct comparison with other results, the 2θ values of the obtained diffractograms were recalculated to the ones of Cu-$K_{\alpha 1}$ radiation. See also [195].

E.2 Time-of-Flight Secondary Ion Mass Spectrometry (ToF-SIMS)

The basic principle of *Secondary Ion Mass Spectrometry* (SIMS) is the bombardment of a sample surface of interest with high-energy so-called primary (or probing) ions and the subsequent detection of ejected secondary ions by mass analysis. Thereby, information about chemical composition of the surface or within several micrometers depth can be gathered.[353]

One option to achieve mass spectrometry is the time-of-flight principle (ToF-SIMS). The secondary ions are ejected using a short-pulsed primary ion gun to reduce temporal variations of the point in time when the ions leave the sample surface.[353, 489] Afterwards, the ejected ions are accelerated by a defined electric field towards the mass analyzer. Based on the flight time in an adjacent flight path with fixed length, the mass/charge ratios of all secondary ions can be determined simultaneously.[353, 489]

Using ToF-SIMS, the chemical composition of samples can be analyzed with large dynamic range of detectable amounts of ions and high mass resolution, e.g., enabling the detection of different isotopes.[353] In particular, the formation of clusters of different ions is characteristic for high-energy sputtering which are detected in ToF-SIMS data.[353] However, especially the interpretation of depth profiles can be challenging since the sputter time of a specific layer depends on the material properties and sputter source. Moreover, additional side effects of sputtering such as the topography can influence the measured signals.[353] See, e.g., [353, 489] for further details.

Experimental Instrumentation Used in This Work

The samples were measured with a *ToF-SIMS 5* setup from *ION-TOF GmbH* with negative polarity (*negative* SIMS). Pulsed primary ions from a 30 keV Bi^+ liquid-metal ion gun were utilized as analytical source, and a second 1 keV Cs source was applied for sputtering. The actual ToF-SIMS analysis was performed on a $100\,\mu m \times 100\,\mu m$ area inside the $300\,\mu m \times 300\,\mu m$ sputtering crater. Measurements were conducted by Jonas Hanisch at the *Zentrum für Sonnenenergie- und Wasserstoff-Forschung Baden-Württemberg* (ZSW) in Stuttgart.

Appendix F

Samples Investigated In This Work

Within this work, both perovskite thin films on glass and fully-processed working thin-film solar cells with perovskite absorber layer have been investigated and are listed in the following.

Perovskite Thin Films on Glass Substrates

All thin-film samples were fabricated by Nadja Giesbrecht, Philipp Angloher, or Meltem F. Aygüler at the *Ludwig-Maximilian University* (LMU) in Munich using standard spin-coating of a precursor solution and subsequent annealing. See, e.g., [238, 239, 241] for further details. The compositions are derived from the initial precursor solutions and might differ in the final prepared films.

Temperature-Dependent Absorption Spectroscopy

The thin-film samples for temperature-dependent absorption spectroscopy (see chapters 7.2 and 8) were capped with a thin layer of PMMA to prevent degradation.

Sample	Composition	Comment
ML0010	$MAPbI_3$	$x = 0.0$
ML0012	$MAPbBr_3$	$x = 1.0$
ML0017	$MAPb(I_{0.5}Br_{0.5})_3$	$x = 0.5$
ML0018	$MAPb(I_{0.75}Br_{0.25})_3$	$x = 0.25$
ML0020	$Cs_{0.1}FA_{0.765}MA_{0.135}Pb(I_{0.765}Br_{0.235})_3$	Cs0.10
ML0025	$MAPb(I_{0.25}Br_{0.75})_3$	$x = 0.75$
ML0031	$Cs_{0.05}(FA_{0.83}MA_{0.17})_{0.95}Pb(I_{0.83}Br_{0.17})_3$	Cs0.05

Table F.1: Perovskite thin films on glass substrates with different composition investigated by temperature-dependent absorption spectroscopy (see chapters 7.2 and 8).

Temperature-Dependent X-Ray Diffraction

The samples for temperature-dependent X-ray powder diffraction (see chapter 8.3) were obtained by scraping off thin films from their glass substrate as mentioned in appendix E.1.

Sample	Composition	Comment
ML0100	$MAPbI_3$	$x = 0.0$
ML0101	$MAPbBr_3$	$x = 1.0$
FMA289	$Cs_{0.05}(FA_{0.83}MA_{0.17})_{0.95}Pb(I_{0.83}Br_{0.17})_3$	Cs0.05

Table F.2: Perovskite samples prepared for temperature-dependent X-ray powder diffraction (see chapter 8.3).

Perovskite Solar-Cell Samples

Hysteresis in j–V Scans

The sample used as exemplary demonstration of dynamic hysteresis effects in j–V scans (see Fig. 2.9) was prepared by Tobias Abzieher at the *Light Technology Institute* (LTI) at *KIT*. It was fabricated by spin-coating and subsequent annealing using a standard layer stack. See, e.g., [324] for further details. Similar samples have been studied in more detail in collaboration with Nico Weber and are presented in his bachelor's thesis.[267]

Sample	Absorber	Layer Stack
C276	$MAPbI_3$	Glass/FTO/TiO_2/$MAPbI_3$/Spiro-MeOTAD/Au

Table F.3: Exemplary sample demonstrating dynamic hysteresis effects in j–V characteristics (see Fig. 2.9). See [267] for more detailed studies.

Establishing Electromodulation Spectroscopy on Perovskite Solar Cells

The samples used for first measurements to establish electromodulation spectroscopy on perovskite solar cells (see chapter 6) were prepared by Jonas Schwenzer at the *Light Technology Institute* (LTI) at *KIT*. They were fabricated based on typical layer stacks by standard spin-coating and subsequent annealing. See, e.g., [324] for further details.

In order to prevent a current flow through the device, the hole-transporting material Spiro-OMeTAD was replaced by a thin PMMA layer. Further investigations of similar samples can be found in the master's thesis of Ina Kelz.[388]

Sample	Absorber	Layer Stack / Comment
300	$MAPbI_3$	Glass/ITO/TiO_2/$MAPbI_3$/PMMA(230 nm)/Au Modified sample with PMMA, SEM image (Fig. 6.6)
306	$MAPbI_3$	Glass/ITO/TiO_2/$MAPbI_3$/PMMA(230 nm)/Au Diffuse electroreflectance measurements (Fig. 6.3)
I312	$MAPbI_3$	Glass/ITO/SnO_2/$MAPbI_3$/Spiro-MeOTAD/Au Comparison with PMMA-modified sample (Fig. 6.6)

Table F.4: Modified samples with insulating PMMA layer instead of spiro-OMeTAD and standard reference sample. for establishing electromodulation spectroscopy on perovskite solar cells (see chapter 6). See [388] for further measurements.

Further standard solar cells with $MAPbI_3$ or $Cs_{0.05}(FA_{0.83}MA_{0.17})_{0.95}Pb(I_{0.83}Br_{0.17})_3$ (Cs0.05) absorber layer were fabricated by Meltem F. Aygüler at the *Ludwig-Maximilian University* (LMU) in Munich. The commonly used layer stack glass/FTO/ TiO_2/perovskite/ Spiro-MeOTAD/Au was used. See, e.g., [241] for further details. Additional investigations of similar samples can be found in [388, 389].

Sample	Absorber	Comment
FMA188-04	$MAPbI_3$	Frequency dependence of $\Delta R/R$ spectra (Fig. 6.8)
FMA201-02	$MAPbI_3$	Voltage dependence of $\Delta R/R$ spectra (Fig. 6.8)
FMA211-03	$MAPbI_3$	Angular dependence of $\Delta R/R$ spectra (Fig. 6.3) and determ. of suitable mod. voltage (Fig. 6.6)
FMA274-04	Cs0.05	Modulus transformation (Fig. 5.6)
FMA288-05	Cs0.05	SEM image (Fig. 2.8 (b))

Table F.5: Standard solar-cell samples with $MAPbI_3$ or $Cs_{0.05}(FA_{0.83}MA_{0.17})_{0.95}$ $Pb(I_{0.83}Br_{0.17})_3$ (Cs0.05) absorber layer in a glass/FTO/TiO_2/perovskite/Spiro-MeOTAD/Au layer stack for establishing electromodulation spectroscopy on perovskite solar cells (see chapter 6). See [388, 389] for further results.

Temperature-Dependent Electromodulation Spectroscopy

For temperature-dependent electroreflectance spectroscopy investigations (see Fig. 8.4), further standard perovskite solar cells with $MAPbI_3$ or $Cs_{0.05}(FA_{0.83}MA_{0.17})_{0.95}Pb(I_{0.83}Br_{0.17})_3$ (Cs0.05) in a glass/FTO/TiO_2/perovskite/ Spiro-MeOTAD/Au layer stack fabricated by Meltem F. Aygüler at the *Ludwig- Maximilian University* (LMU) in Munich have been used (see above). Additional results of similar samples can be found in [389].

Sample	Absorber	
FMA201-03	$MAPbI_3$	
FMA201-06	Cs0.05	$Cs_{0.05}(FA_{0.83}MA_{0.17})_{0.95}Pb(I_{0.83}Br_{0.17})_3$
FMA221-03	Cs0.05	$Cs_{0.05}(FA_{0.83}MA_{0.17})_{0.95}Pb(I_{0.83}Br_{0.17})_3$

Table F.6: Standard solar-cell samples with glass/FTO/TiO_2/perovskite/ Spiro-MeOTAD/Au layer stack used for temperature-dependent electroreflectance spectroscopy investigations (see Fig. 8.4). See [389] for additional results.

In order to compare electromodulation spectra and standard absorption spectra (see chapters 6.2, 7.1, 7.3, 8.1 and appendix C), semitransparent solar-cells with transparent contact layers are necessary. These devices were prepared by Moritz Schultes at the *Zentrum für Sonnenenergie- und Wasserstoff-Forschung Baden-Württemberg* (ZSW) in Stuttgart using sputtered IZO[223] with an additional thin MoO_3 protection layer (see Fig. 2.8). See [389] for further results on similar samples.

Sample	Absorber	Layer Stack
P367-FV140	$MAPbI_3$	Glass/FTO/TiO_2/PCBM/$MAPbI_3$/ Spiro-MeOTAD/MoO_3/IZO
P376-FV235	$MAPbI_3$	Glass/FTO/TiO_2/PCBM/$MAPbI_3$/ Spiro-MeOTAD/MoO_3/IZO
ZSW-S01 (for SEM image in Fig. 2.8 (c))	$MAPbI_3$	Glass/FTO/TiO_2/PCBM/$MAPbI_3$/ Spiro-MeOTAD/MoO_3/IZO

Table F.7: Semitransparent perovskite solar cells for comparison of electromodulation and standard absorption spectra (see chapters 6.2, 7.1, 7.3, 8.1 and appendix C). See [389] for further results on similar samples.

Temperature-Dependent Photocurrent Spectroscopy

The semitransparent perovskite solar cells used for temperature-dependent photocurrent spectroscopy (see chapter 7.3 and appendix A.2) were fabricated by Moritz Schultes at the *Zentrum für Sonnenenergie- und Wasserstoff-Forschung Baden-Württemberg* (ZSW) in Stuttgart (see above, here with additional SiO_2 nanoparticles for improved layer quality[240]) and Ihteaz Hossain at the *Light Technology Institute* (LTI) at *KIT* (here with sputtered ITO as second transparent contact layer). Further measurement results have been obtained in collaboration with Philip Lange and are presented in his master's thesis.[422]

Sample	Absorber	Layer Stack
261d (ZSW)	$MAPbI_3$	Glass/ITO/SnO_2/C60-SAM/SiO_2-NP/ $MAPbI_3$/Spiro-MeOTAD/MoO_3/IZO
B116 (LTI)	$Cs_{0.1}(FA_{0.83}MA_{0.17})_{0.9}$ $Pb(I_{0.83}Br_{0.17})_3$	Glass/ITO/SnO_2/$Cs_{0.1}(FA_{0.83}MA_{0.17})_{0.9}$ $Pb(I_{0.83}Br_{0.17})_3$/Spiro-MeOTAD/ITO

Table F.8: Semitransparent perovskite solar cells for temperature-dependent photocurrent spectroscopy (see chapter 7.3 and appendix A.2). Additional results on similar samples can be found in [422].

Electroreflectance Studies of Bandgap Instabilities

All samples used for the investigations of reversible bandgap instabilities in multiple-cation mixed halide perovskite solar cells (see chapter 9) were prepared by Meltem F. Aygüler at the *Ludwig- Maximilian University* (LMU) in Munich using the standard Glass/FTO/ TiO_2/perovskite/Spiro-MeOTAD/Au layer stack (see above). The used sample layout consists of 12 cells per sample which allows for different investigations of the same sample, e.g., time-of-flight secondary ion mass spectrometry (ToF-SIMS) and electroreflectance spectroscopy. The standard absorber composition $Cs_{0.05}(FA_{0.83}MA_{0.17})_{0.95}Pb(I_{0.83}Br_{0.17})_3$ is abbreviated as $(Cs,FA,MA)Pb(I,Br)_3$. $(Cs,FA,MA)PbI_3$ and $(FA,MA)PbI_3$ similarly denote $Cs_{0.05}(FA_{0.83}MA_{0.17})_{0.95}PbI_3$ and $FA_{0.83}MA_{0.17}PbI_3$, respectively.

Figure F.1 presents typical values of the power-conversion efficiency of these standard perovskite solar cells with different absorber compositions.

Sample	Absorber	Comment
FMA211-04	$(Cs,FA,MA)Pb(I,Br)_3$	Comparison of compositions (Fig. 9.2 (a)), atmospheric conditions (Fig. 9.10), spectra at high RH (Figs. 9.8 + 9.9)
FMA211-10	$(FA,MA)PbI_3$	Comparison of compositions (Fig. 9.2 (a))
FMA211-14	$(Cs,FA,MA)PbI_3$	Comparison of compositions (Fig. 9.2 (a)) line-width broadening (Fig. 9.2 (b)), spectra at high RH (Fig. 9.9)
FMA221-02	$MAPbI_3$	Comparison of compositions (Fig. 9.2 (a))
FMA221-03	$(Cs,FA,MA)Pb(I,Br)_3$	Check of reversibility (Fig. 9.5), and atmospheric conditions (Fig. 9.10)
FMA221-04	$(Cs,FA,MA)Pb(I,Br)_3$	Check of reversibility (Fig. 9.5), and atmospheric conditions (Fig. 9.10)
FMA253-08	$(Cs,FA,MA)Pb(I,Br)_3$	In-situ X-Ray Diffraction (Fig. 9.3)
FMA274-01	$(Cs,FA,MA)Pb(I,Br)_3$	Comparison of atmospheric conditions (Fig. 9.10)
FMA274-02	$(Cs,FA,MA)Pb(I,Br)_3$	Comparison of atmospheric conditions (Fig. 9.10), influence of dc bias (Fig. 9.7)
FMA274-04	$(Cs,FA,MA)Pb(I,Br)_3$	Shift in $\Delta R/R$ Spectra (Fig. 9.1), comparison of broadening (Fig. 9.2 (b)), influence of dc bias (Fig. 9.7), and ToF-SIMS (Fig. 9.4)
FMA274-05	$(Cs,FA,MA)Pb(I,Br)_3$	Influence of dc bias (Fig. 9.7)

Table F.9: Standard perovskite solar-cell samples with glass/FTO/TiO_2/perovskite/Spiro-MeOTAD/Au layer stack used for the investigations of reversible bandgap instabilities (see chapter 9).

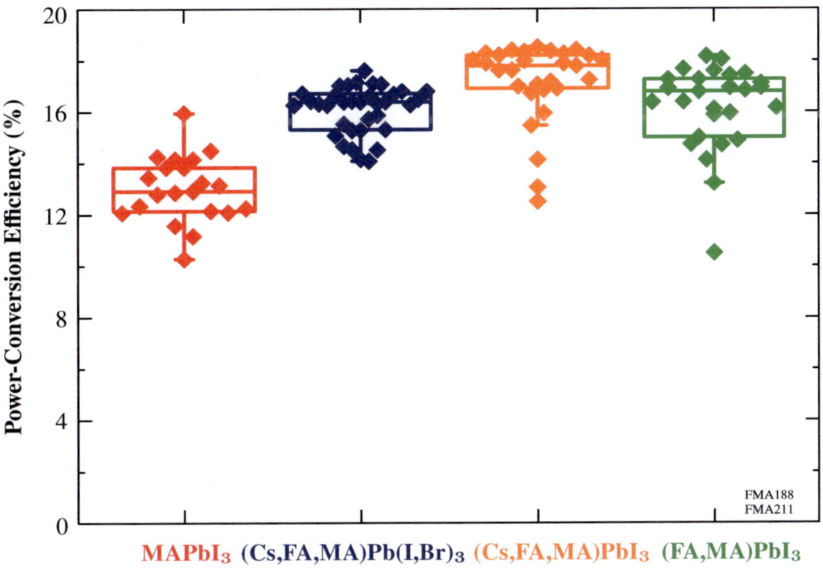

Figure F.1: Typical values of power-conversion efficiency (PCE) of standard perovskite solar cells with different composition of the perovskite absorber which were fabricated by Meltem F. Aygüler at the *Ludwig- Maximilian University* (LMU) in Munich. Mean PCE values between 12% and 19% demonstrate the good quality of perovskite absorber layers and solar cells. Reprinted with permission from [241]. Copyright 2018 American Chemical Society.

List of Publications

Regular Articles

1. **Excitonic Nature of Optical Transitions in Electroabsorption Spectra of Perovskite Solar Cells,**
   F. Ruf, A. Magin, M. Schultes, E. Ahlswede, H. Kalt, and M. Hetterich,
   Applied Physics Letters **112**, 083902 (2018) - **Featured Article**

   Published Press Releases:

   - AIP Scilight:
     Perovskite Solar Cells Reveal Excitonic Optical Transitions
     `https://aip.scitation.org/doi/10.1063/1.5026230` (2018/02/19)
   - KIT Press Release 035/2018:
     Why Perovskite Solar Cells Are So Efficient
     `https://www.kit.edu/kit/english/pi_2018_035_why-perovskite-solar-cells-are-so-efficient.php` (2018/04/04)
   - pv magazine:
     KIT und Oxford PV machen Fortschritte in der Perowskit-Forschung
     `https://www.pv-magazine.de/2018/04/04/kit-und-oxford-pv-machen-fortschritte-in-der-perowskit-forschung/` (2018/04/04)
   - German Ministry of Education and Research (BMBF) :
     Was Perowskit-Solarzellen so effizient macht
     `https://www.fona.de/de/was-perowskit-solarzellen-so-effizient-macht-23347.html` (2018/05/15)

2. **The Bandgap as a Moving Target: Reversible Bandgap Instabilities in Multiple-Cation Mixed-Halide Perovskite Solar Cells,**
   F. Ruf, P. Rietz, M. F. Aygüler, I. Kelz, P. Docampo, H. Kalt, and M. Hetterich,
   ACS Energy Letters **3** (12), 2995–3001 (2018)

3. **Temperature-Dependent Studies of Exciton Binding Energy and Phase-Transition Suppression in (Cs,FA,MA)Pb(I,Br)$_3$ Perovskites,**
   F. Ruf, M. F. Aygüler, N. Giesbrecht, B. Rendenbach, A. Magin, P. Docampo, H. Kalt, and M. Hetterich, *APL Materials* **7**, 031113 (2019)

Contributions to International Conferences

1. **Temperature-Dependent Spectroscopy and Exciton Binding Energy of Mixed-Cation Lead Mixed-Halide Perovskite Absorber Materials**,
F. Ruf, N. Giesbrecht, P. Angloher, P. Docampo, H. Kalt, and M. Hetterich,
2nd International Conference on Perovskite Solar Cells and Optoelectronics (PSCO), Genova, Italy (2016), Poster Presentation

2. **Temperature-Dependent Optical Spectroscopy and Determination of Exciton Binding Energy of Mixed-Cation Lead Mixed-Halide Perovskite Absorber Materials for Tandem Solar Cells**,
F. Ruf, N. Giesbrecht, P. Angloher, P. Docampo, H. Kalt, and M. Hetterich,
MRS Fall Meeting and Exhibit, Boston, MA, USA (2016), Poster Presentation

3. **Electroreflectance Spectroscopy on Organic–Inorganic Halide Perovskite Solar Cells**,
F. Ruf, I. Kelz, A. Magin, J. Schwenzer, T. Abzieher, S. Moghadamzadeh, M. F. Aygüler, P. Docampo, H. Kalt, and M. Hetterich, *3rd International Conference on Perovskite Solar Cells and Optoelectronics (PSCO), Oxford, United Kingdom (2017)*, Poster Presentation

4. **Temperature-Dependent Electromodulation Spectroscopy of Excitons in Perovskite Solar Cells**,
F. Ruf, A. Magin, M. Schultes, M. F. Aygüler, P. Docampo, E. Ahlswede, H. Kalt, and M. Hetterich, *World Conference on Photovoltaic Energy Conversion (WCPEC-7), Waikoloa, HI, USA (2018)*, Oral Presentation and Proceedings Article

5. **Reversible Bandgap Instabilities in Multiple-Cation Mixed-Halide Perovskite Solar Cells**,
F. Ruf, P. Rietz, M. F. Aygüler, P. Docampo, H. Kalt, and M. Hetterich, *International Conference on Perovskite Solar Cells, Photonics and Optoelectronics (NIPHO19), Jerusalem, Israel (2019)*, Oral Presentation and Proceedings Article

Contributions to Workshops and Seminars

1. **Temperature-Dependent Exciton Binding Energy and Transient Photocurrent in Perovskite Absorber Layers and Solar Cells**,
F. Ruf, N. Giesbrecht, N. Weber, P. Angloher, T. Abzieher, P. Docampo, H. Kalt, and M. Hetterich, *BMBF Workshop Perovskites, Würzburg, Germany (2016)*, Poster Presentation

2. **Optical Spectroscopy of Perovskite Structures for Thin-Film Solar Cells**,
F. Ruf, *Karlsruhe Days of Optics and Photonics (KDOP), Karlsruhe, Germany (2017)*, Poster Presentation

3. **Electromodulation Spectroscopy on Organic–Inorganic Perovskite Solar Cells — From Excitons to Bandgap Instabilities,**
F. Ruf, *Photovoltaics Day at KIT, Karlsruhe, Germany (2018)*, Invited Oral Presentation

Contributions to DPG Spring Meetings

1. **Temperature-Dependent Absorption Measurements and Photoluminescence Spectroscopy of Mixed Organic–Inorganic Halide Perovskite films,**
A. Schulz, F. Ruf, N. Giesbrecht, M. Handloser, P. Docampo, H. Kalt, and M. Hetterich, *DPG Spring Meeting, Regensburg (2016)*, Poster Presentation

2. **Temperature-Dependent Transient Photocurrent Measurements on Organic–Inorganic Halide Perovskite Solar Cells,**
I. Kelz, N. Weber, F. Ruf, T. Abzieher, N. Giesbrecht, M. F. Aygüler, P. Docampo, H. Kalt, and M. Hetterich, *DPG Spring Meeting, Dresden (2017)*, Poster Presentation

3. **Electromodulation Spectroscopy on Organic–Inorganic Perovskite Solar Cells,**
A. Magin, F. Ruf, M. Schultes, E. Ahlswede, J. Schwenzer, H. Kalt, and M. Hetterich, *DPG Spring Meeting, Berlin (2018)*, Poster Presentation

4. **Temperature-Dependent Electroabsorption Spectroscopy on Organic–Inorganic Perovskite Solar Cells,**
F. Ruf, A. Magin, M. Schultes, E. Ahlswede, H. Kalt, and M. Hetterich, *DPG Spring Meeting, Berlin (2018)*, Oral Presentation

5. **Reversible Changes of the Bandgap Energy in Multiple-Cation Mixed-Halide Perovskite Solar Cells Under Illumination and Bias Investigated by Optical Spectroscopy,**
E. Wirth, F. Ruf, M. F. Aygüler, J. Hanisch, P. Docampo, E. Ahlswede, H. Kalt, and M. Hetterich, *DPG Spring Meeting, Regensburg (2019)*, Poster Presentation

6. **Simulation of Electroreflectance Spectra of CIGS Solar Cells,**
E. Förstner, A. Magin, F. Ruf, C. Huber, W. Witte, D. Hariskos, O. Kiowski, H. Kalt, and M. Hetterich, *DPG Spring Meeting, Regensburg (2019)*, Poster Presentation

7. **Influence of Excitonic Effects on Charge-Carrier Extraction in Organic–Inorganic Perovskite Solar Cells,**
P. Lange, F. Ruf, I. M. Hossain, U. W. Paetzold, M. Schultes, E. Ahlswede, H. Kalt, and M. Hetterich, *DPG Spring Meeting, Regensburg (2019)*, Poster Presentation

8. **Reversible Bandgap Instabilities in Multiple-Cation Mixed-Halide Perovskite Solar Cells,**
F. Ruf, P. Rietz, M. F. Aygüler, P. Docampo, H. Kalt, and M. Hetterich, *DPG Spring Meeting, Regensburg (2019)*, Oral Presentation

Bibliography

[1] P. Würfel, *Physics of Solar Cells: From Basic Principles to Advanced Concepts* (Wiley-VCH, Weinheim, 2009), 2nd ed.

[2] B. Parida, S. Iniyan, and R. Goic, "A review of solar photovoltaic technologies," Renewable and Sustainable Energy Reviews **15**, 1625–1636 (2011).

[3] BP, "BP Statistical Review of World Energy 2019," `https://www.bp.com/content/dam/bp/business-sites/en/global/corporate/pdfs/energy-economics/statistical-review/bp-stats-review-2019-full-report.pdf`, accessed 2019-06-16.

[4] N. Mejriszan Toosi, J. A. Esfahani, and M. Reza Safaei, "Investigation of energy consumption and renewable energy resources in top ten countries with most energy consumption," IOP Conference Series: Materials Science and Engineering **397**, 012108 (2018).

[5] A. E. Becquerel, "Memoire sur les effects electriques produits sous l'influence des rayons solaires," Comptes Rendus de L'Academie des Sciences **9**, 561–567 (1839).

[6] L. El Chaar, L. A. Lamont, and N. El Zein, "Review of photovoltaic technologies," Renewable and Sustainable Energy Reviews **15**, 2165–2175 (2011).

[7] International Renewable Energy Agency, "Global Energy Transformation: A Roadmap to 2050," `https://irena.org/-/media/Files/IRENA/Agency/Publication/2018/Apr/IRENA_Report_GET_2018.pdf`, accessed 2019-06-15.

[8] BP, "BP Energy Outlook 2019," `https://www.bp.com/content/dam/bp/business-sites/en/global/corporate/pdfs/energy-economics/energy-outlook/bp-energy-outlook-2019.pdf`, accessed 2019-06-15.

[9] Fraunhofer Institute for Solar Energy Systems, "Photovoltaics Report," `https://www.ise.fraunhofer.de/content/dam/ise/de/documents/publications/studies/Photovoltaics-Report.pdf`, accessed 2019-06-15.

[10] National Renewable Research Laboratory, "NREL Efficiency Chart," `http://www.nrel.gov/pv/`, accessed 2019-06-15.

[11] M. A. Green, J. Zhao, A. Wang, and S. R. Wenham, "Progress and outlook for high-efficiency crystalline silicon solar cells," Solar Energy Materials and Solar Cells **65**, 9–16 (2001).

[12] S. Pizzini, "Towards solar grade silicon: Challenges and benefits for low cost photovoltaics," Solar Energy Materials and Solar Cells **94**, 1528–1533 (2010).

[13] T. D. Lee and A. U. Ebong, "A review of thin film solar cell technologies and challenges," Renewable and Sustainable Energy Reviews **70**, 1286–1297 (2017).

[14] M. Nakamura, "Record cell 1 cm^2-sized Cd-free CIGSSe," presented at "10th International Workshop on CIGS Solar Cell Technology (IW-CIGSTech-10)," (Paris, France, 2019).

[15] C. B. Carter and M. G. Norton, *Ceramic Materials: Science and Engineering* (Springer, New York, NY, 2007).

[16] H.-R. Wenk and A. Bulakh, *Minerals: Their Constitution and Origin* (Cambridge University Press, Cambridge, 2004).

[17] A. S. Bhalla, R. Guo, and R. Roy, "The perovskite structure – a review of its role in ceramic science and technology," Materials Research Innovations **4**, 3–26 (2000).

[18] M. Johnsson and P. Lemmens, "Perovskites and thin films – crystallography and chemistry," Journal of Physics: Condensed Matter **20**, 264001 (2008).

[19] H. J. Snaith, "Perovskites: The Emergence of a New Era for Low-Cost, High-Efficiency Solar Cells," Journal of Physical Chemistry Letters **4**, 3623–3630 (2013).

[20] M. A. Green, A. Ho-Baillie, and H. J. Snaith, "The emergence of perovskite solar cells," Nature Photonics **8**, 506–514 (2014).

[21] T. C. Sum and N. Mathews, "Advancements in perovskite solar cells: photophysics behind the photovoltaics," Energy and Environmental Science **7**, 2518–2534 (2014).

[22] J. M. Frost and A. Walsh, "Molecular Motion and Dynamic Crystal Structures of Hybrid Halide Perovskites," in "Organic–Inorganic Halide Perovskite Photovoltaics: From Fundamentals to Device Architectures," , N. G. Park, M. Grätzel, and T. Miyasaka, eds. (Springer International Publishing, Switzerland, 2016), chap. 1, pp. 1–18.

[23] D. Weber, "$CH_3NH_3PbX_3$, ein Pb(II)-System mit kubischer Perowskitstruktur," Zeitschrift für Naturforschung **33 b**, 1443–1445 (1978).

[24] A. Kojima, K. Teshima, Y. Shirai, and T. Miyasaka, "Organometal Halide Perovskites as Visible-Light Sensitizers for Photovoltaic Cells," Journal of the American Chemical Society **131**, 6050–6051 (2009).

[25] H. S. Kim, C. R. Lee, J. H. Im, K. B. Lee, T. Moehl, A. Marchioro, S. J. Moon, R. Humphry-Baker, J. H. Yum, J. E. Moser, M. Grätzel, and N. G. Park, "Lead Iodide Perovskite Sensitized All-Solid-State Submicron Thin Film Mesoscopic Solar Cell with Efficiency Exceeding 9%," Scientific Reports **2**, 591 (2012).

[26] M. M. Lee, J. Teuscher, T. Miyasaka, T. N. Murakami, and H. J. Snaith, "Efficient Hybrid Solar Cells Based on Meso-Superstructured Organometal Halide Perovskites," Science **338**, 643–647 (2012).

[27] J. M. Ball, M. M. Lee, A. Hey, and H. J. Snaith, "Low-temperature processed meso-superstructured to thin-film perovskite solar cells," Energy and Environmental Science **6**, 1739–1743 (2013).

[28] M. Liu, M. B. Johnston, and H. J. Snaith, "Efficient planar heterojunction perovskite solar cells by vapour deposition," Nature **501**, 395–398 (2013).

[29] M. L. Petrus, J. Schlipf, C. Li, T. P. Gujar, N. Giesbrecht, P. Müller-Buschbaum, M. Thelakkat, T. Bein, S. Hüttner, and P. Docampo, "Capturing the Sun: A Review of the Challenges and Perspectives of Perovskite Solar Cells," Advanced Energy Materials **7**, 1700264 (2017).

[30] Y. Zhao and K. Zhu, "Organic–inorganic hybrid lead halide perovskites for optoelectronic and electronic applications," Chemical Society Reviews **45**, 655–689 (2016).

[31] Y.-S. Jung, K. Hwang, Y.-J. Heo, J.-E. Kim, D. Vak, and D.-Y. Kim, "Progress in Scalable Coating and Roll-to-Roll Compatible Printing Processes of Perovskite Solar Cells toward Realization of Commercialization," Advanced Optical Materials **6**, 1701182 (2018).

[32] Clarivate Analytics, "Web of Science," `http://apps.webofknowledge.com`, accessed 2019-06-01.

[33] M. Era, S. Morimoto, T. Tsutsui, and S. Saito, "Organic-inorganic heterostructure electroluminescent device using a layered perovskite semiconductor $(C_6H_5C_2H_4NH_3)_2PbI_4$," Applied Physics Letters **65**, 676–678 (1994).

[34] Z. K. Tan, R. S. Moghaddam, M. L. Lai, P. Docampo, R. Higler, F. Deschler, M. Price, A. Sadhanala, L. M. Pazos, D. Credgington, F. Hanusch, T. Bein, H. J. Snaith, and R. H. Friend, "Bright light-emitting diodes based on organometal halide perovskite," Nature Nanotechnology **9**, 687–692 (2014).

[35] H. P. Kim, J. Kim, B. S. Kim, H.-M. Kim, J. Kim, A. R. b. M. Yusoff, J. Jang, and M. K. Nazeeruddin, "High-Efficiency, Blue, Green, and Near-Infrared Light-Emitting Diodes Based on Triple Cation Perovskite," Advanced Optical Materials **5**, 1600920 (2017).

[36] Y. Cao, N. Wang, H. Tian, J. Guo, Y. Wei, H. Chen, Y. Miao, W. Zou, K. Pan, Y. He, H. Cao, Y. Ke, M. Xu, Y. Wang, M. Yang, K. Du, Z. Fu, D. Kong, D. Dai, Y. Jin, G. Li, H. Li, Q. Peng, J. Wang, and W. Huang, "Perovskite light–emitting diodes based on spontaneously formed submicrometre–scale structures," Nature **562**, 249–253 (2018).

[37] K. Lin, J. Xing, L. N. Quan, F. P. G. de Arquer, X. Gong, J. Lu, L. Xie, W. Zhao, D. Zhang, C. Yan, W. Li, X. Liu, Y. Lu, J. Kirman, E. H. Sargent, Q. Xiong, and Z. Wei, "Perovskite light-emitting diodes with external quantum efficiency exceeding 20 per cent," Nature **562**, 245–248 (2018).

[38] M. Saliba, S. M. Wood, J. B. Patel, P. K. Nayak, J. Huang, J. A. Alexander-Webber, B. Wenger, S. D. Stranks, M. T. Hörantner, J. T. W. Wang, R. J. Nicholas, L. M. Herz, M. B. Johnston, S. M. Morris, H. J. Snaith, and M. K. Riede, "Structured Organic–Inorganic Perovskite toward a Distributed Feedback Laser," Advanced Materials **28**, 923–929 (2016).

[39] Y. Jia, R. A. Kerner, A. J. Grede, B. P. Rand, and N. C. Giebink, "Continuous-wave lasing in an organic–inorganic lead halide perovskite semiconductor," Nature Photonics **11**, 784–788 (2017).

[40] P. Brenner, T. Glöckler, D. Rueda-Delgado, T. Abzieher, M. Jakoby, B. S. Richards, U. W. Paetzold, I. A. Howard, and U. Lemmer, "Triple cation mixed-halide perovskites for tunable lasers," Optical Materials Express **7**, 4082 (2017).

[41] F. Mathies, P. Brenner, G. Hernandez-Sosa, I. A. Howard, U. W. Paetzold, and U. Lemmer, "Inkjet-printed perovskite distributed feedback lasers," Optics Express **26**, A144–A152 (2018).

[42] P. Brenner, O. Bar-On, M. Jakoby, I. Allegro, B. S. Richards, U. W. Paetzold, I. A. Howard, J. Scheuer, and U. Lemmer, "Continuous wave amplified spontaneous emission in phase-stable lead halide perovskites," Nature Communications **10**, 988 (2019).

[43] Q. Zhang, S. T. Ha, X. Liu, T. C. Sum, and Q. Xiong, "Room-Temperature Near-Infrared High-Q Perovskite Whispering-Gallery Planar Nanolasers," Nano Letters **14**, 5995–6001 (2014).

[44] B. R. Sutherland, S. Hoogland, M. M. Adachi, C. T. O. Wong, and E. H. Sargent, "Conformal Organohalide Perovskites Enable Lasing on Spherical Resonators," ACS Nano **8**, 10947–10952 (2014).

[45] J. Zhao, Y. Yan, C. Wei, W. Zhang, Z. Gao, and Y. S. Zhao, "Switchable Single-Mode Perovskite Microlasers Modulated by Responsive Organic Microdisks," Nano Letters **18**, 1241–1245 (2018).

[46] L. Dou, Y. M. Yang, J. You, Z. Hong, W. H. Chang, G. Li, and Y. Yang, "Solution-processed hybrid perovskite photodetectors with high detectivity," Nature Communications **5**, 5404 (2014).

[47] X. Hu, X. Zhang, L. Liang, J. Bao, S. Li, W. Yang, and Y. Xie, "High-Performance Flexible Broadband Photodetector Based on Organolead Halide Perovskite," Advanced Functional Materials **24**, 7373–7380 (2014).

[48] R. Dong, Y. Fang, J. Chae, J. Dai, Z. Xiao, Q. Dong, Y. Yuan, A. Centrone, X. C. Zeng, and J. Huang, "High-Gain and Low-Driving-Voltage Photodetectors Based on Organolead Triiodide Perovskites," Advanced Materials **27**, 1912–1918 (2015).

[49] P. V. Kamat and C. M. Biegel, "Perovskite Stories from Around the World," ACS Energy Letters **4**, 879–887 (2019).

[50] M. I. Asghar, J. Zhang, H. Wang, and P. D. Lund, "Device stability of perovskite solar cells – A review," Renewable and Sustainable Energy Reviews **77**, 131–146 (2017).

[51] Q. Fu, X. Tang, B. Huang, T. Hu, L. Tan, L. Chen, and Y. Chen, "Recent Progress on the Long-Term Stability of Perovskite Solar Cells," Advanced Science **5**, 1700387 (2018).

[52] C. C. Boyd, R. Cheacharoen, T. Leijtens, and M. D. McGehee, "Understanding Degradation Mechanisms and Improving Stability of Perovskite Photovoltaics," Chemical Reviews **119**, 3418–3451 (2019).

[53] H. Cho, Y.-H. Kim, C. Wolf, H.-D. Lee, and T.-W. Lee, "Improving the Stability of Metal Halide Perovskite Materials and Light-Emitting Diodes," Advanced Materials **30**, 1704587 (2018).

[54] G. Grancini, C. Roldán-Carmona, I. Zimmermann, E. Mosconi, X. Lee, D. Martineau, S. Narbey, F. Oswald, F. De Angelis, M. Graetzel, and M. K. Nazeeruddin, "One-Year stable perovskite solar cells by 2D/3D interface engineering," Nature Communications **8**, 15684 (2017).

[55] P. Holzhey and M. Saliba, "A full overview of international standards assessing the long-term stability of perovskite solar cells," Journal of Materials Chemistry A **6**, 21794–21808 (2018).

[56] C. C. Boyd, R. Cheacharoen, K. A. Bush, R. Prasanna, T. Leijtens, and M. D. McGehee, "Barrier Design to Prevent Metal-Induced Degradation and Improve Thermal Stability in Perovskite Solar Cells," ACS Energy Letters **3**, 1772–1778 (2018).

[57] K. A. Bush, A. F. Palmstrom, Z. J. Yu, M. Boccard, R. Cheacharoen, J. P. Mailoa, D. P. McMeekin, R. L. Z. Hoye, C. D. Bailie, T. Leijtens, I. M. Peters, M. C. Minichetti, N. Rolston, R. Prasanna, S. Sofia, D. Harwood, W. Ma, F. Moghadam, H. J. Snaith, T. Buonassisi, Z. C. Holman, S. F. Bent, and M. D. McGehee, "23.6%-efficient monolithic perovskite/silicon tandem solar cells with improved stability," Nature Energy **2**, 17009 (2017).

[58] L. Shi, T. L. Young, J. Kim, Y. Sheng, L. Wang, Y. Chen, Z. Feng, M. J. Keevers, X. Hao, P. J. Verlinden, M. A. Green, and A. W. Ho-Baillie, "Accelerated Lifetime Testing of Organic–Inorganic Perovskite Solar Cells Encapsulated by Polyisobutylene," ACS Applied Materials and Interfaces **9**, 25073–25081 (2017).

[59] B. O. Seraphin and N. Bottka, "Band-Structure Analysis from Electro-Reflectance Studies," Physical Review **145**, 628–636 (1966).

[60] A. G. Thompson, M. Cardona, and K. L. Shacklee, "Electroreflectance in the GaAs–GaP Alloys," Physical Review **146**, 601–610 (1966).

[61] M. Cardona, K. L. Shaklee, and F. H. Pollak, "Electroreflectance at a Semiconductor–Electrolyte Interface," Physical Review **154**, 696–720 (1967).

[62] F. H. Pollak and M. Cardona, "Piezo-Electroreflectance in Ge, GaAs, and Si," Physical Review **172**, 816–837 (1968).

[63] F. Cerdeira and M. Cardona, "Photoreflectance and Electroreflectance in Silicon," Solid State Communications **7**, 879–882 (1969).

[64] J. L. Shay, "Photoreflectance Line Shape at the Fundamental Edge in Ultrapure GaAs," Physical Review B **2**, 803–807 (1970).

[65] D. E. Aspnes and A. A. Studna, "Schottky-Barrier Electroreflectance: Application to GaAs," Physical Review B **7**, 4605–4625 (1973).

[66] S. Tachi, A. Moritani, and J. Nakai, "Alloy Composition Profile of Heterogeneous Structure in $GaAs_{1-x}P_x$ Determined by the Anodization-Electroreflectance Technique," Journal of Electrochemical Society **124**, 1948–1949 (1977).

[67] E. E. Mendez, L. L. Chang, G. Landgren, R. Ludeke, L. Esaki, and F. H. Pollak, "Observation of Superlattice Effects on the Electronic Bands of Multilayer Heterostructures," Physical Review Letters **46**, 1230–1234 (1981).

[68] M. Erman, J. B. Theeten, P. Frijlink, S. Gaillard, F. J. Hia, and C. Alibert, "Electronic states and thicknesses of GaAs/GaAlAs quantum wells as measured by electroreflectance and spectroscopic ellipsometry," Journal of Applied Physics **56**, 3241–3249 (1984).

[69] C. Alibert, S. Gaillard, J. A. Brum, G. Bastard, P. Frijlink, and M. Erman, "Measurements of Electric-Field-Induced Energy-Level Shifts in GaAs Single-Quantum-Wells Using Electroreflectance," Solid State Communications **53**, 457–460 (1985).

[70] O. J. Glembocki, B. V. Shanabrook, N. Bottka, W. T. Beard, and J. Comas, "Photoreflectance characterization of interband transitions in GaAs/AlGaAs multiple quantum wells and modulation-doped heterojunctions," Applied Physics Letters **46**, 970–972 (1985).

[71] O. J. Glembocki, B. V. Shanabrook, N. Bottka, W. T. Beard, and J. Comas, "Photoreflectance Characterization Of GaAs/AlGaAs Thin Films, Heterojunctions And Multiple Quantum Well Structures," Proceedings of SPIE **524** (1985).

[72] M. Galluppi, L. Geelhaar, H. Riechert, M. Hetterich, A. Grau, S. Birner, and W. Stolz, "Bound-to-bound and bound-to-free transitions in surface photovoltage spectra: Determination of the band offsets for $In_xGa_{1-x}As$ and $In_xGa_{1-x}As_{1-y}N_y$ quantum wells," Physical Review B **72**, 155324 (2005).

[73] A. Grau, T. Passow, and M. Hetterich, "Temperature dependence of the GaAsN conduction band structure," Applied Physics Letters **89**, 202105 (2006).

[74] C. Huber, C. Krämmer, D. Sperber, A. Magin, H. Kalt, and M. Hetterich, "Electroreflectance of thin-film solar cells: Simulation and experiment," Physical Review B **92**, 75201 (2015).

[75] C. Krämmer, C. Huber, A. Redinger, D. Sperber, G. Rey, S. Siebentritt, H. Kalt, and M. Hetterich, "Diffuse electroreflectance of thin-film solar cells: Suppression of interference-related lineshape distortions," Applied Physics Letters **107**, 222104 (2015).

[76] F. Ruf, A. Magin, M. Schultes, E. Ahlswede, H. Kalt, and M. Hetterich, "Excitonic nature of optical transitions in electroabsorption spectra of perovskite solar cells," Applied Physics Letters **112**, 083902 (2018).

[77] R. J. Elliott, "Intensity of Optical Absorption by Excitons," Physical Review **108**, 1384–1389 (1957).

[78] H. Haug and S. W. Koch, *Quantum Theory of the Optical and Electronic Properties of Semiconductors* (World Scientific, Singapore, 1993), 2nd ed.

[79] M. Saba, M. Cadelano, D. Marongiu, F. Chen, V. Sarritzu, N. Sestu, C. Figus, M. Aresti, R. Piras, A. Geddo Lehmann, C. Cannas, A. Musinu, F. Quochi, A. Mura, and G. Bongiovanni, "Correlated electron–hole plasma in organometal perovskites," Nature Communications **5**, 5049 (2014).

[80] A. Poglitsch and D. Weber, "Dynamic disorder in methylammoniumtrihalogeno-plumbates (II) observed by millimeter-wave spectroscopy," Journal of Chemical Physics **87**, 6373–6378 (1987).

[81] M. A. Green, Y. Jiang, A. M. Soufiani, and A. Ho-Baillie, "Optical Properties of Photovoltaic Organic–Inorganic Lead Halide Perovskites," Journal of Physical Chemistry Letters **6**, 4774–4785 (2015).

[82] M. T. Weller, O. J. Weber, P. F. Henry, A. M. Di Pumpo, and T. C. Hansen, "Complete structure and cation orientation in the perovskite photovoltaic methylammonium lead iodide between 100 and 352 K," Chemical Communications **51**, 4180–4183 (2015).

[83] F. Brivio, J. M. Frost, J. M. Skelton, A. J. Jackson, O. J. Weber, M. T. Weller, A. R. Goñi, A. M. Leguy, P. R. Barnes, and A. Walsh, "Lattice dynamics and vibrational spectra of the orthorhombic, tetragonal, and cubic phases of methylammonium lead iodide," Physical Review B **92**, 144308 (2015).

[84] P. S. Whitfield, N. Herron, W. E. Guise, K. Page, Y. Q. Cheng, I. Milas, and M. K. Crawford, "Structures, Phase Transitions and Tricritical Behavior of the Hybrid Perovskite Methyl Ammonium Lead Iodide," Scientific Reports **6**, 35685 (2016).

[85] C. Quarti, E. Mosconi, and F. De Angelis, "Interplay of Orientational Order and Electronic Structure in Methylammonium Lead Iodide: Implications for Solar Cells Operation," Chemistry of Materials **26**, 6557–6569 (2014).

[86] C. C. Stoumpos, C. D. Malliakas, and M. G. Kanatzidis, "Semiconducting Tin and Lead Iodide Perovskites with Organic Cations: Phase Transitions, High Mobilities, and Near-Infrared Photoluminescent Properties," Inorganic Chemistry **52**, 9019–9038 (2013).

[87] W. Geng, L. Zhang, Y. N. Zhang, W. M. Lau, and L. M. Liu, "First-Principles Study of Lead Iodide Perovskite Tetragonal and Orthorhombic Phases for Photovoltaics," Journal of Physical Chemistry C **118**, 19565–19571 (2014).

[88] S. T. A. G. Melissen, F. Labat, P. Sautet, and T. Le Bahers, "Electronic properties of $PbX_3CH_3NH_3$ (X = Cl, Br, I) compounds for photovoltaic and photocatalytic applications," Physical Chemistry Chemical Physics **17**, 2199–2209 (2015).

[89] H. H. Fang, R. Raissa, M. Abdu-Aguye, S. Adjokatse, G. R. Blake, J. Even, and M. A. Loi, "Photophysics of Organic–Inorganic Hybrid Lead Iodide Perovskite Single Crystals," Advanced Functional Materials **25**, 2378–2385 (2015).

[90] I. Deretzis, B. N. Di Mauro, A. Alberti, G. Pellegrino, E. Smecca, and A. La Magna, "Spontaneous bidirectional ordering of $CH_3NH_3^+$ in lead iodide perovskites at room temperature: The origins of the tetragonal phase," Scientific Reports **6**, 24443 (2016).

[91] T. Umebayashi, K. Asai, T. Kondo, and A. Nakao, "Electronic structures of lead iodide based low-dimensional crystals," Physical Review B **67**, 155405 (2003).

[92] G. Giorgi, J. I. Fujisawa, H. Segawa, and K. Yamashita, "Small Photocarrier Effective Masses Featuring Ambipolar Transport in Methylammonium Lead Iodide Perovskite: A Density Functional Analysis," Journal of Physical Chemistry Letters **4**, 4213–4216 (2013).

[93] A. Filippetti and A. Mattoni, "Hybrid perovskites for photovoltaics: Insights from first principles," Physical Review B **89**, 125203 (2014).

[94] R. A. Jishi, O. B. Ta, and A. A. Sharif, "Modeling of Lead Halide Perovskites for Photovoltaic Applications," Journal of Physical Chemistry C **118**, 28344–28349 (2014).

[95] T. Ahmed, C. La-O-Vorakiat, T. Salim, Y. M. Lam, E. E. Chia, and J. X. Zhu, "Optical properties of organometallic perovskite: An ab initio study using relativistic GW correction and Bethe-Salpeter equation," EPL (Europhysics Letters) **108**, 67015 (2014).

[96] A. Amat, E. Mosconi, E. Ronca, C. Quarti, P. Umari, M. K. Nazeeruddin, M. Grätzel, and F. De Angelis, "Cation-Induced Band-Gap Tuning in Organohalide Perovskites: Interplay of Spin–Orbit Coupling and Octahedra Tilting," Nano Letters **14**, 3608–3616 (2014).

[97] E. Menéndez-Proupin, P. Palacios, P. Wahnón, and J. C. Conesa, "Self-consistent relativistic band structure of the $CH_3NH_3PbI_3$ perovskite," Physical Review B **90**, 045207 (2014).

[98] J. Even, L. Pedesseau, and C. Katan, "Analysis of Multivalley and Multibandgap Absorption and Enhancement of Free Carriers Related to Exciton Screening in Hybrid Perovskites," Journal of Physical Chemistry C **118**, 11566–11572 (2014).

[99] J. Even, L. Pedesseau, C. Katan, M. Kepenekian, J. S. Lauret, D. Sapori, and E. Deleporte, "Solid-State Physics Perspective on Hybrid Perovskite Semiconductors," Journal of Physical Chemistry C **119**, 10161–10177 (2015).

[100] S. X. Tao, X. Cao, and P. A. Bobbert, "Accurate and efficient band gap predictions of metal halide perovskites using the DFT-1/2 method: GW accuracy with DFT expense," Scientific Reports **7**, 14386 (2017).

[101] W. Yin, J. Yang, J. Kang, Y. Yan, and S.-H. Wei, "Halide perovskite materials for solar cells: a theoretical review," Journal of Materials Chemistry A **3**, 8926–8942 (2015).

[102] M. R. Filip, G. E. Eperon, H. J. Snaith, and F. Giustino, "Steric engineering of metal–halide perovskites with tunable optical band gaps," Nature Communications **5**, 5757 (2014).

[103] M. Grätzel, "The Rise of Highly Efficient and Stable Perovskite Solar Cells," Accounts of Chemical Research **50**, 487–491 (2017).

[104] R. E. Brandt, V. Stevanović, D. S. Ginley, and T. Buonassisi, "Identifying defect-tolerant semiconductors with high minority-carrier lifetimes: beyond hybrid lead halide perovskites," MRS Communications **5**, 265–275 (2015).

[105] R. E. Brandt, J. R. Poindexter, P. Gorai, R. C. Kurchin, R. L. Hoye, L. Nienhaus, M. W. Wilson, J. A. Polizzotti, R. Sereika, R. Žaltauskas, L. C. Lee, J. L. Macmanus-Driscoll, M. Bawendi, V. Stevanović, and T. Buonassisi, "Searching for 'Defect-Tolerant' Photovoltaic Materials: Combined Theoretical and Experimental Screening," Chemistry of Materials **29**, 4667–4674 (2017).

[106] M. V. Kovalenko, L. Protesescu, and M. I. Bodnarchuk, "Properties and potential optoelectronic applications of lead halide perovskite nanocrystals," Science **358**, 745–750 (2017).

[107] A. M. Ganose, C. N. Savory, and D. O. Scanlon, "Beyond methylammonium lead iodide: prospects for the emergent field of ns^2 containing solar absorbers," Chemical Communications **53**, 20–44 (2017).

[108] J. Kim, S. H. Lee, J. H. Lee, and K. H. Hong, "The Role of Intrinsic Defects in Methylammonium Lead Iodide Perovskite," Journal of Physical Chemistry Letters **5**, 1312–1317 (2014).

[109] M. H. Du, “Efficient carrier transport in halide perovskites: theoretical perspectives,” Journal of Materials Chemistry A **2**, 9091–9098 (2014).

[110] J. M. Azpiroz, E. Mosconi, J. Bisquert, and F. De Angelis, “Defect migration in methylammonium lead iodide and its role in perovskite solar cell operation,” Energy and Environmental Science **8**, 2118–2127 (2015).

[111] W.-J. Yin, T. Shi, and Y. Yan, “Unusual defect physics in $CH_3NH_3PbI_3$ perovskite solar cell absorber,” Applied Physics Letters **104**, 063903 (2014).

[112] Y. Yan, W.-J. Yin, T. Shi, W. Meng, and C. Feng, “Defect Physics of $CH_3NH_3PbX_3$ (X = I, Br, Cl) Perovskites,” in “Organic–Inorganic Halide Perovskite Photovoltaics: From Fundamentals to Device Architectures,” , N. G. Park, M. Grätzel, and T. Miyasaka, eds. (Springer International Publishing, Switzerland, 2016), chap. 4, pp. 79–106.

[113] W. Ming, S. Chen, and M. H. Du, “Chemical instability leads to unusual chemical-potential- independent defect formation and diffusion in perovskite solar cell material $CH_3NH_3PbI_3$,” Journal of Materials Chemistry A **4**, 16975–16981 (2016).

[114] D. Yang, W. Ming, H. Shi, L. Zhang, and M. H. Du, “Fast Diffusion of Native Defects and Impurities in Perovskite Solar Cell Material $CH_3NH_3PbI_3$,” Chemistry of Materials **28**, 4349–4357 (2016).

[115] A. Buin, P. Pietsch, J. Xu, O. Voznyy, A. H. Ip, R. Comin, and E. H. Sargent, “Materials Processing Routes to Trap-Free Halide Perovskites,” Nano Letters **14**, 6281–6286 (2014).

[116] D. Meggiolaro, S. G. Motti, E. Mosconi, A. J. Barker, J. Ball, C. A. R. Perini, F. Deschler, A. Petrozza, and F. De Angelis, “Iodine chemistry determines the defect tolerance of lead–halide perovskites,” Energy and Environmental Science **11**, 702–713 (2018).

[117] M. L. Agiorgousis, Y.-Y. Sun, H. Zeng, and S. Zhang, “Strong Covalency-Induced Recombination Centers in Perovskite Solar Cell Material $CH_3NH_3PbI_3$,” Journal of the American Chemical Society **136**, 14570–14575 (2014).

[118] H. S. Duan, H. Zhou, Q. Chen, P. Sun, S. Luo, T. B. Song, B. Bob, and Y. Yang, “The identification and characterization of defect states in hybrid organic–inorganic perovskite photovoltaics,” Physical Chemistry Chemical Physics **17**, 112–116 (2015).

[119] M. Samiee, S. Konduri, B. Ganapathy, R. Kottokkaran, H. A. Abbas, A. Kitahara, P. Joshi, L. Zhang, M. Noack, and V. Dalal, “Defect density and dielectric constant in perovskite solar cells,” Applied Physics Letters **105**, 153502 (2014).

[120] Y. Hu, T. Bein, P. Docampo, M. F. Aygüler, A. G. Hufnagel, T. J. Savenije, V. Dyakonov, A. Hartschuh, K. Tvingstedt, J. Hanisch, P. Rieder, I. Grill, M. Handloser, M. L. Petrus, A. Baumann, and E. M. Hutter, "Understanding the Role of Cesium and Rubidium Additives in Perovskite Solar Cells: Trap States, Charge Transport, and Recombination," Advanced Energy Materials **8**, 1703057 (2018).

[121] A. Baumann, S. Väth, P. Rieder, M. C. Heiber, K. Tvingstedt, and V. Dyakonov, "Identification of Trap States in Perovskite Solar Cells," Journal of Physical Chemistry Letters **6**, 2350–2354 (2015).

[122] A. Cuevas, M. Stocks, D. McDonald, M. Kerr, and C. Samundsett, "Recombination and Trapping in Multicrystalline Silicon," IEEE Transactions on Electron Devices **46**, 2026–2034 (1999).

[123] P. Gundel, M. C. Schubert, and W. Warta, "Origin of trapping in multicrystalline silicon," Journal of Applied Physics **104**, 073716 (2008).

[124] C. Eames, J. M. Frost, P. R. Barnes, B. C. O'Regan, A. Walsh, and M. S. Islam, "Ionic transport in hybrid lead iodide perovskite solar cells," Nature Communications **6**, 7497 (2015).

[125] Y. Yuan and J. Huang, "Ion Migration in Organometal Trihalide Perovskite and Its Impact on Photovoltaic Efficiency and Stability," Accounts of Chemical Research **49**, 286–293 (2016).

[126] P. Delugas, C. Caddeo, A. Filippetti, and A. Mattoni, "Thermally Activated Point Defect Diffusion in Methylammonium Lead Trihalide: Anisotropic and Ultrahigh Mobility of Iodine," Journal of Physical Chemistry Letters **7**, 2356–2361 (2016).

[127] A. Senocrate, I. Moudrakovski, G. Y. Kim, T. Y. Yang, G. Gregori, M. Grätzel, and J. Maier, "The Nature of Ion Conduction in Methylammonium Lead Iodide: A Multimethod Approach," Angewandte Chemie – International Edition **56**, 7755–7759 (2017).

[128] Y. Yuan, J. Chae, Y. Shao, Q. Wang, Z. Xiao, A. Centrone, and J. Huang, "Photovoltaic Switching Mechanism in Lateral Structure Hybrid Perovskite Solar Cells," Advanced Energy Materials **5**, 1500615 (2015).

[129] C. Li, S. Tscheuschner, F. Paulus, P. E. Hopkinson, J. Kießling, A. Köhler, Y. Vaynzof, and S. Huettner, "Iodine Migration and its Effect on Hysteresis in Perovskite Solar Cells," Advanced Materials **28**, 2446–2454 (2016).

[130] D. Meggiolaro, E. Mosconi, and F. De Angelis, "Formation of Surface Defects Dominates Ion Migration in Lead-Halide Perovskites," ACS Energy Letters **4**, 779–785 (2019).

[131] K. Domanski, J. P. Correa-Baena, N. Mine, M. K. Nazeeruddin, A. Abate, M. Saliba, W. Tress, A. Hagfeldt, and M. Grätzel, "Not All That Glitters Is Gold: Metal-Migration-Induced Degradation in Perovskite Solar Cells," ACS Nano **10**, 6306–6314 (2016).

[132] V. M. Goldschmidt, "Die Gesetze der Krystallochemie," Die Naturwissenschaften **14**, 477–485 (1926).

[133] C. Li, X. Lu, W. Ding, L. Feng, Y. Gao, and Z. Guo, "Formability of ABX_3 (X = F, Cl, Br, I) halide perovskites," Acta Crystallographica Section B **64**, 702–707 (2008).

[134] H. S. Kim, S. H. Im, and N. G. Park, "Organolead Halide Perovskite: New Horizons in Solar Cell Research," Journal of Physical Chemistry C **118**, 5615–5625 (2014).

[135] G. Kieslich, S. Sun, and A. K. Cheetham, "Solid-state principles applied to organic–inorganic perovskites: new tricks for an old dog," Chemical Science **5**, 4712–4715 (2014).

[136] C. Li, K. C. K. Soh, and P. Wu, "Formability of ABO_3 perovskites," Journal of Alloys and Compounds **372**, 40–48 (2004).

[137] B. N. Cohen, C. Labarca, N. Davidson, and H. A. Lester, "Mutations in M2 Alter the Selectivity of the Mouse Nicotinic Acetylcholine Receptor for Organic and Alkali Metal Cations," Journal of General Physiology **100**, 373–400 (1992).

[138] M. Saliba, J. P. Correa-Baena, M. Grätzel, A. Hagfeldt, and A. Abate, "Perovskite Solar Cells: From the Atomic Level to Film Quality and Device Performance," Angewandte Chemie – International Edition **57**, 2554–2569 (2018).

[139] C. K. Møller, "Crystal Structure and Photoconductivity of Cæsium Plumbohalides," Nature **182**, 1436 (1958).

[140] J. W. Xiao, L. Liu, D. Zhang, N. De Marco, J. W. Lee, O. Lin, Q. Chen, and Y. Yang, "The Emergence of the Mixed Perovskites and Their Applications as Solar Cells," Advanced Energy Materials **7**, 1700491 (2017).

[141] J. W. Lee, D. H. Kim, H. S. Kim, S. W. Seo, S. M. Cho, and N. G. Park, "Formamidinium and Cesium Hybridization for Photo- and Moisture-Stable Perovskite Solar Cell," Advanced Energy Materials **5**, 1501310 (2015).

[142] Z. Li, M. Yang, J. S. Park, S. H. Wei, J. J. Berry, and K. Zhu, "Stabilizing Perovskite Structures by Tuning Tolerance Factor: Formation of Formamidinium and Cesium Lead Iodide Solid-State Alloys," Chemistry of Materials **28**, 284–292 (2016).

[143] D. P. McMeekin, G. Sadoughi, W. Rehman, G. E. Eperon, M. Saliba, M. T. Horantner, A. Haghighirad, N. Sakai, L. Korte, B. Rech, M. B. Johnston, L. M. Herz, and H. J. Snaith, "A mixed-cation lead mixed-halide perovskite absorber for tandem solar cells," Science **351**, 151–155 (2016).

[144] D. P. McMeekin, Z. Wang, W. Rehman, F. Pulvirenti, J. B. Patel, N. K. Noel, M. B. Johnston, S. R. Marder, L. M. Herz, and H. J. Snaith, "Crystallization Kinetics and Morphology Control of Formamidinium–Cesium Mixed-Cation Lead Mixed-Halide Perovskite via Tunability of the Colloidal Precursor Solution," Advanced Materials **29**, 1607039 (2017).

[145] W. Shockley and H. J. Queisser, "Detailed Balance Limit of Efficiency of p–n Junction Solar Cells," Journal of Applied Physics **32**, 510–519 (1961).

[146] S. Rühle, "Tabulated values of the Shockley-Queisser limit for single junction solar cells," Solar Energy **130**, 139–147 (2016).

[147] N. Pellet, P. Gao, G. Gregori, T. Yang, M. K. Nazeeruddin, J. Maier, and M. Grätzel, "Mixed-Organic-Cation Perovskite Photovoltaics for Enhanced Solar-Light Harvesting," Angewandte Chemie – International Edition **53**, 3151–3157 (2014).

[148] T. Duong, Y. Wu, H. Shen, J. Peng, X. Fu, D. Jacobs, E.-C. Wang, T. C. Kho, K. C. Fong, M. Stocks, E. Franklin, A. Blakers, N. Zin, K. McIntosh, W. Li, Y.-B. Cheng, T. P. White, K. Weber, and K. Catchpole, "Rubidium Multication Perovskite with Optimized Bandgap for Perovskite-Silicon Tandem with over 26% Efficiency," Advanced Energy Materials **7**, 1700228 (2017).

[149] Y. Wu, D. Yan, J. Peng, T. Duong, Y. Wan, S. P. Phang, H. Shen, N. Wu, C. Barugkin, X. Fu, S. Surve, D. Grant, D. Walter, T. P. White, K. R. Catchpole, and K. J. Weber, "Monolithic perovskite/silicon-homojunction tandem solar cell with over 22% efficiency," Energy and Environmental Science **10**, 2472–2479 (2017).

[150] F. Sahli, J. Werner, B. A. Kamino, M. Bräuninger, R. Monnard, B. Paviet-Salomon, L. Barraud, L. Ding, J. J. Diaz Leon, D. Sacchetto, G. Cattaneo, M. Despeisse, M. Boccard, S. Nicolay, Q. Jeangros, B. Niesen, and C. Ballif, "Fully textured monolithic perovskite/silicon tandem solar cells with 25.2% power conversion efficiency," Nature Materials **17**, 820–826 (2018).

[151] U. W. Paetzold, M. Jaysankar, R. Gehlhaar, E. Ahlswede, S. Paetel, W. Qiu, J. Bastos, L. Rakocevic, B. S. Richards, T. Aernouts, M. Powalla, and J. Poortmans, "Scalable perovskite/CIGS thin-film solar module with power conversion efficiency of 17.8%," Journal of Materials Chemistry A **5**, 9897–9906 (2017).

[152] Q. Han, Y.-T. Hsieh, L. Meng, J.-L. Wu, P. Sun, E.-P. Yao, S.-Y. Chang, S.-H. Bae, T. Kato, V. Bermudez, and Y. Yang, "High-performance perovskite/Cu(In,Ga)Se_2 monolithic tandem solar cells," Science **361**, 904–908 (2018).

[153] H. Shen, T. Duong, J. Peng, D. Jacobs, N. Wu, J. Gong, Y. Wu, S. K. Karuturi, X. Fu, K. Weber, X. Xiao, T. P. White, and K. Catchpole, "Mechanically-stacked perovskite/CIGS tandem solar cells with efficiency of 23.9% and reduced oxygen sensitivity," Energy and Environmental Science **11**, 394–406 (2018).

[154] M. Jaysankar, M. Filipič, B. Zielinski, R. Schmager, W. Song, W. Qiu, U. W. Paetzold, T. Aernouts, M. Debucquoy, R. Gehlhaar, and J. Poortmans, "Perovskite–silicon tandem solar modules with optimised light harvesting," Energy and Environmental Science **11**, 1489–1498 (2018).

[155] N. N. Lal, Y. Dkhissi, W. Li, Q. Hou, Y. B. Cheng, and U. Bach, "Perovskite Tandem Solar Cells," Advanced Energy Materials **7**, 1602761 (2017).

[156] M. T. Hörantner and H. J. Snaith, "Predicting and optimising the energy yield of perovskite-on-silicon tandem solar cells under real world conditions," Energy and Environmental Science **10**, 1983–1993 (2017).

[157] C. M. Sutter-Fella, Y. Li, M. Amani, J. W. Ager, F. M. Toma, E. Yablonovitch, I. D. Sharp, and A. Javey, "High Photoluminescence Quantum Yield in Band Gap Tunable Bromide Containing Mixed Halide Perovskites," Nano Letters **16**, 800–806 (2016).

[158] B. W. Park, B. Philippe, S. M. Jain, X. Zhang, T. Edvinsson, H. Rensmo, B. Zietz, and G. Boschloo, "Chemical engineering of methylammonium lead iodide/bromide perovskites: Tuning of opto-electronic properties and photovoltaic performance," Journal of Materials Chemistry A **3**, 21760–21771 (2015).

[159] L. Gil-Escrig, A. Miquel-Sempere, M. Sessolo, and H. J. Bolink, "Mixed Iodide–Bromide Methylammonium Lead Perovskite-based Diodes for Light Emission and Photovoltaics," Journal of Physical Chemistry Letters **6**, 3743–3748 (2015).

[160] Z. Xiao, R. A. Kerner, N. Tran, L. Zhao, G. D. Scholes, and B. P. Rand, "Engineering Perovskite Nanocrystal Surface Termination for Light-Emitting Diodes with External Quantum Efficiency Exceeding 15%," Advanced Functional Materials **29**, 1807284 (2019).

[161] G. E. Eperon, T. Leijtens, K. A. Bush, R. Prasanna, T. Green, J. T.-W. Wang, D. P. McMeekin, G. Volonakis, R. L. Milot, R. May, A. Palmstrom, D. J. Slotcavage, R. A. Belisle, J. B. Patel, E. S. Parrott, R. J. Sutton, W. Ma, F. Moghadam, B. Conings, A. Babayigit, H.-G. Boyen, S. Bent, F. Giustino, L. M. Herz, M. B. Johnston, M. D. McGehee, and H. J. Snaith, "Perovskite–perovskite tandem photovoltaics with optimized band gaps," Science **354**, 861–865 (2016).

[162] A. Rajagopal, Z. Yang, S. B. Jo, I. L. Braly, P. W. Liang, H. W. Hillhouse, and A. K. Jen, "Highly Efficient Perovskite–Perovskite Tandem Solar Cells Reaching 80% of the Theoretical Limit in Photovoltage," Advanced Materials **29**, 1702140 (2017).

[163] T. Leijtens, R. Prasanna, K. A. Bush, G. E. Eperon, J. A. Raiford, A. Gold-Parker, E. J. Wolf, S. A. Swifter, C. C. Boyd, H. P. Wang, M. F. Toney, S. F. Bent, and M. D. McGehee, "Tin–lead halide perovskites with improved thermal and air stability for efficient all-perovskite tandem solar cells," Sustainable Energy and Fuels **2**, 2450–2459 (2018).

[164] D. Zhao, Y. Yu, C. Wang, W. Liao, N. Shrestha, C. R. Grice, A. J. Cimaroli, L. Guan, R. J. Ellingson, K. Zhu, X. Zhao, R. G. Xiong, and Y. Yan, "Low-bandgap mixed tin–lead iodide perovskite absorbers with long carrier lifetimes for all-perovskite tandem solar cells," Nature Energy **2**, 17018 (2017).

[165] D. Forgács, L. Gil-Escrig, D. Pérez-Del-Rey, C. Momblona, J. Werner, B. Niesen, C. Ballif, M. Sessolo, and H. J. Bolink, "Efficient Monolithic Perovskite/Perovskite Tandem Solar Cells," Advanced Energy Materials **7**, 1602121 (2017).

[166] A. F. Palmstrom, G. E. Eperon, T. Leijtens, R. Prasanna, S. N. Habisreutinger, W. Nemeth, E. A. Gaulding, S. P. Dunfield, M. Reese, S. Nanayakkara, T. Moot, J. Werner, J. Liu, B. To, S. T. Christensen, M. D. McGehee, M. F. van Hest, J. M. Luther, J. J. Berry, and D. T. Moore, "Enabling Flexible All-Perovskite Tandem Solar Cells," Joule **3**, 1–12 (2019).

[167] M. Saliba, T. Matsui, J.-Y. Seo, K. Domanski, J.-P. Correa-Baena, M. K. Nazeeruddin, S. M. Zakeeruddin, W. Tress, A. Abate, A. Hagfeldt, and M. Grätzel, "Cesium-containing triple cation perovskite solar cells: improved stability, reproducibility and high efficiency," Energy and Environmental Science **9**, 1989–1997 (2016).

[168] M. Saliba, T. Matsui, K. Domanski, J.-Y. Seo, A. Ummadisingu, S. M. Zakeeruddin, J.-P. Correa-Baena, W. R. Tress, A. Abate, A. Hagfeldt, and M. Grätzel, "Incorporation of rubidium cations into perovskite solar cells improves photovoltaic performance," Science **354**, 206–209 (2016).

[169] T. Matsui, T. Yokoyama, T. Negami, T. Sekiguchi, M. Saliba, M. Grätzel, and H. Segawa, "Effect of Rubidium for Thermal Stability of Triple-cation Perovskite Solar Cells," Chemistry Letters **47**, 814–816 (2018).

[170] Y. H. Park, I. Jeong, S. Bae, H. J. Son, P. Lee, J. Lee, C. H. Lee, and M. J. Ko, "Inorganic Rubidium Cation as an Enhancer for Photovoltaic Performance and Moisture Stability of $HC(NH_2)_2PbI_3$ Perovskite Solar Cells," Advanced Functional Materials **27**, 1605988 (2017).

[171] W. Tan, A. R. Bowring, A. C. Meng, M. D. McGehee, and P. C. McIntyre, "Thermal Stability of Mixed Cation Metal Halide Perovskites in Air," ACS Applied Materials and Interfaces **10**, 5485–5491 (2018).

[172] Y. Hu, M. F. Aygüler, M. L. Petrus, T. Bein, and P. Docampo, "Impact of Rubidium and Cesium Cations on the Moisture Stability of Multiple-Cation Mixed-Halide Perovskites," ACS Energy Letters **2**, 2212–2218 (2017).

[173] F. Zheng, W. Chen, T. Bu, K. P. Ghiggino, F. Huang, Y. Cheng, P. Tapping, T. W. Kee, B. Jia, and X. Wen, "Triggering the Passivation Effect of Potassium Doping in Mixed-Cation Mixed-Halide Perovskite by Light Illumination," Advanced Energy Materials **9**, 1901016 (2019).

[174] M. Abdi-Jalebi, Z. Andaji-Garmaroudi, S. Cacovich, C. Stavrakas, B. Philippe, J. M. Richter, M. Alsari, E. P. Booker, E. M. Hutter, A. J. Pearson, S. Lilliu, T. J. Savenije, H. Rensmo, G. Divitini, C. Ducati, R. H. Friend, and S. D. Stranks, "Maximizing and stabilizing luminescence from halide perovskites with potassium passivation," Nature **555**, 497–501 (2018).

[175] Y. Sheng, A. Mei, S. Liu, M. Duan, P. Jiang, C. Tian, Y. Xiong, Y. Rong, H. Han, and Y. Hu, "Mixed $(5\text{-AVA})_x MA_{1-x} PbI_{3-y}(BF_4)_y$ perovskites enhance the photovoltaic performance of hole-conductor-free printable mesoscopic solar cells," Journal of Materials Chemistry A **6**, 2360–2364 (2018).

[176] N. D. Marco, H. Zhou, Q. Chen, P. Sun, Z. Liu, L. Meng, E. P. Yao, Y. Liu, A. Schiffer, and Y. Yang, "Guanidinium: A Route to Enhanced Carrier Lifetime and Open-Circuit Voltage in Hybrid Perovskite Solar Cells," Nano Letters **16**, 1009–1016 (2016).

[177] L. K. Ono, E. J. Juarez-Perez, and Y. Qi, "Progress on Perovskite Materials and Solar Cells with Mixed Cations and Halide Anions," ACS Applied Materials and Interfaces **9**, 30197–30246 (2017).

[178] I. C. Smith, E. T. Hoke, D. Solis-Ibarra, M. D. McGehee, and H. I. Karunadasa, "A Layered Hybrid Perovskite Solar-Cell Absorber with Enhanced Moisture Stability," Angewandte Chemie – International Edition **53**, 11232–11235 (2014).

[179] T. M. Koh, K. Thirumal, H. S. Soo, and N. Mathews, "Multidimensional Perovskites: A Mixed Cation Approach Towards Ambient Stable and Tunable Perovskite Photovoltaics," ChemSusChem **9**, 2541–2558 (2016).

[180] Y. Chen, Y. Sun, J. Peng, J. Tang, K. Zheng, and Z. Liang, "2D Ruddlesden–Popper Perovskites for Optoelectronics," Advanced Materials **30**, 1703487 (2018).

[181] D. H. Cao, C. C. Stoumpos, O. K. Farha, J. T. Hupp, and M. G. Kanatzidis, "2D Homologous Perovskites as Light-Absorbing Materials for Solar Cell Applications," Journal of the American Chemical Society **137**, 7843–7850 (2015).

[182] L. Mao, Y. Wu, C. C. Stoumpos, B. Traore, C. Katan, J. Even, M. R. Wasielewski, and M. G. Kanatzidis, "Tunable White-Light Emission in Single-Cation-Templated Three-Layered 2D Perovskites $(CH_3CH_2NH_3)_4Pb_3Br_{10-x}Cl_x$," Journal of the American Chemical Society **139**, 11956–11963 (2017).

[183] Y. Hu, J. Schlipf, M. Wussler, M. L. Petrus, W. Jaegermann, T. Bein, P. Müller-Buschbaum, and P. Docampo, "Hybrid Perovskite/Perovskite Heterojunction Solar Cells," ACS Nano **10**, 5999–6007 (2016).

[184] N. Li, Z. Zhu, C.-C. Chueh, H. Liu, B. Peng, A. Petrone, X. Li, L. Wang, and A. K.-Y. Jen, "Mixed Cation $FA_xPEA_{1-x}PbI_3$ with Enhanced Phase and Ambient Stability toward High-Performance Perovskite Solar Cells," Advanced Energy Materials **7**, 1601307 (2017).

[185] S. Gharibzadeh, B. Abdollahi Nejand, M. Jakoby, T. Abzieher, D. Hauschild, S. Moghadamzadeh, J. A. Schwenzer, P. Brenner, R. Schmager, A. A. Haghighirad, L. Weinhardt, U. Lemmer, B. S. Richards, I. A. Howard, and U. W. Paetzold, "Record Open-Circuit Voltage Wide-Bandgap Perovskite Solar Cells Utilizing 2D/3D Perovskite Heterostructure," Advanced Energy Materials **9**, 1803699 (2019).

[186] P. Li, Y. Zhang, C. Liang, G. Xing, X. Liu, F. Li, X. Liu, X. Hu, G. Shao, and Y. Song, "Phase Pure 2D Perovskite for High-Performance 2D–3D Heterostructured Perovskite Solar Cells," Advanced Materials **30**, 1805323 (2018).

[187] A. Krishna, S. Gottis, M. K. Nazeeruddin, and F. Sauvage, “Mixed Dimensional 2D/3D Hybrid Perovskite Absorbers: The Future of Perovskite Solar Cells?” Advanced Functional Materials **29**, 1806482 (2019).

[188] G. L. Araújo and A. Martí, “Absolute limiting efficiencies for photovoltaic energy conversion,” Solar Energy Materials and Solar Cells **33**, 213–240 (1994).

[189] M. A. Green, “Photovoltaic principles,” Physica E **14**, 11–17 (2002).

[190] J. Bisquert, D. Cahen, G. Hodes, S. Rühle, and A. Zaban, “Physical Chemical Principles of Photovoltaic Conversion with Nanoparticulate, Mesoporous Dye-Sensitized Solar Cells,” Journal of Physical Chemistry B **108**, 8106–8118 (2004).

[191] S. De Wolf, J. Holovsky, S. J. Moon, P. Löper, B. Niesen, M. Ledinsky, F. J. Haug, J. H. Yum, and C. Ballif, “Organometallic Halide Perovskites: Sharp Optical Absorption Edge and Its Relation to Photovoltaic Performance,” Journal of Physical Chemistry Letters **5**, 1035–1039 (2014).

[192] D. Liu, M. K. Gangishetty, and T. L. Kelly, “Effect of $CH_3NH_3PbI_3$ thickness on device efficiency in planar heterojunction perovskite solar cells,” Journal of Materials Chemistry A **2**, 19873–19881 (2014).

[193] D. Yang, X. Zhou, R. Yang, Z. Yang, W. Yu, X. Wang, C. Li, S. Liu, and R. P. Chang, “Surface optimization to eliminate hysteresis for record efficiency planar perovskite solar cells,” Energy and Environmental Science **9**, 3071–3078 (2016).

[194] W. S. Yang, B.-W. Park, E. H. Jung, N. J. Jeon, Y. C. Kim, D. U. Lee, S. S. Shin, J. Seo, E. K. Kim, J. H. Noh, and S. I. Seok, “Iodide management in formamidinium-lead-halide–based perovskite layers for efficient solar cells,” Science **356**, 1376–1379 (2017).

[195] F. Ruf, M. F. Aygüler, N. Giesbrecht, B. Rendenbach, A. Magin, P. Docampo, H. Kalt, and M. Hetterich, “Temperature-Dependent Studies of Exciton Binding Energy and Phase-Transition Suppression in $(Cs,FA,MA)Pb(I,Br)_3$ Perovskites,” APL Materials **7**, 031113 (2019).

[196] H. Hoppe and N. S. Sariciftci, “Organic solar cells: An overview,” Journal of Materials Research **19**, 1924–1945 (2004).

[197] A. Hagfeldt, U. B. Cappel, G. Boschloo, L. Sun, L. Kloo, H. Petterson, and E. A. Gibson, “Dye-Sensitized and Organic Solar Cells,” in “Solar Cells: Materials, Manufacture and Operation,” , A. McEvoy, L. Castañer, and T. Markvart, eds. (Academic Press, Oxford, 2012), 2nd ed.

[198] W. Tress, “Maximum Efficiency and Open-Circuit Voltage of Perovskite Solar Cells,” in “Organic–Inorganic Halide Perovskite Photovoltaics: From Fundamentals to Device Architectures,” , N. G. Park, M. Grätzel, and T. Miyasaka, eds. (Springer International Publishing, Switzerland, 2016), chap. 3, pp. 53–77.

[199] U. Würfel, A. Cuevas, and P. Würfel, "Charge Carrier Separation in Solar Cells," IEEE Journal of Photovoltaics **5**, 461–469 (2015).

[200] A. Kojima, K. Teshima, T. Miyasaka, and Y. Shirai, "Novel Photoelectrochemical Cell with Mesoscopic Electrodes Sensitized by Lead-halide Compounds," in "210th ECS Meeting," (Cancun, Mexico, 2006), Oct. 29–Nov. 3, 2006.

[201] A. Kojima, K. Teshima, Y. Shirai, and T. Miyasaka, "Novel Photoelectrochemical Cell with Mesoscopic Electrodes Sensitized by Lead-halide Compounds," in "214th ECS Meeting," (Honolulu, HI, USA, 2008), Oct. 12–17, 2008.

[202] M. Grätzel, "The light and shade of perovskite solar cells," Nature Materials **13**, 838–842 (2014).

[203] Y. Li, W. Yan, Y. Li, S. Wang, W. Wang, Z. Bian, L. Xiao, and Q. Gong, "Direct Observation of Long Electron–Hole Diffusion Distance in $CH_3NH_3PbI_3$ Perovskite Thin Film," Scientific Reports **5**, 14485 (2015).

[204] M. B. Johnston and L. M. Herz, "Hybrid Perovskites for Photovoltaics: Charge-Carrier Recombination, Diffusion, and Radiative Efficiencies," Accounts of Chemical Research **49**, 146–154 (2016).

[205] K. Handloser, N. Giesbrecht, T. Bein, P. Docampo, M. Handloser, and A. Hartschuh, "Contactless visualization of fast charge carrier diffusion in hybrid halide perovskite thin films," ACS Photonics **3**, 255–261 (2016).

[206] I. Grill, M. F. Aygüler, T. Bein, P. Docampo, N. F. Hartmann, M. Handloser, and A. Hartschuh, "Charge Transport Limitations in Perovskite Solar Cells: The Effect of Charge Extraction Layers," ACS Applied Materials and Interfaces **9**, 37655–37661 (2017).

[207] V. Gonzalez-Pedro, E. J. Juarez-Perez, W. S. Arsyad, E. M. Barea, F. Fabregat-Santiago, I. Mora-Sero, and J. Bisquert, "General Working Principles of $CH_3NH_3PbX_3$ Perovskite Solar Cells," Nano Letters **14**, 888–893 (2014).

[208] S. Ravishankar, S. Gharibzadeh, C. Roldán-Carmona, G. Grancini, Y. Lee, M. Ralaiarisoa, A. M. Asiri, N. Koch, J. Bisquert, and M. K. Nazeeruddin, "Influence of Charge Transport Layers on Open-Circuit Voltage and Hysteresis in Perovskite Solar Cells," Joule **2**, 788–798 (2018).

[209] T. Kirchartz, J. Bisquert, I. Mora-Sero, and G. Garcia-Belmonte, "Classification of solar cells according to mechanisms of charge separation and charge collection," Physical Chemistry Chemical Physics **17**, 4007–4014 (2015).

[210] A. Guerrero, E. J. Juarez-Perez, J. Bisquert, I. Mora-Sero, and G. Garcia-Belmonte, "Electrical field profile and doping in planar lead halide perovskite solar cells," Applied Physics Letters **105**, 133902 (2014).

[211] J. Bisquert, *The Physics of Solar Cells: Perovskites, Organics, and Photovoltaic Fundamentals* (CRC Press, Taylor & Francis Group, Boca Raton, FL, 2018).

[212] E. Edri, S. Kirmayer, S. Mukhopadhyay, K. Gartsman, G. Hodes, and D. Cahen, "Elucidating the charge carrier separation and working mechanism of $CH_3NH_3PbI_{3-x}Cl_x$ perovskite solar cells," Nature Communications **5**, 3461 (2014).

[213] X. Sun, R. Asadpour, W. Nie, A. D. Mohite, and M. A. Alam, "A Physics-Based Analytical Model for Perovskite Solar Cells," IEEE Journal of Photovoltaics **5**, 1389–1394 (2015).

[214] T. Minemoto and M. Murata, "Device modeling of perovskite solar cells based on structural similarity with thin film inorganic semiconductor solar cells," Journal of Applied Physics **116**, 054505 (2014).

[215] I. Mora-Seró, "How Do Perovskite Solar Cells Work?" Joule **2**, 585–587 (2018).

[216] D. A. Jacobs, Y. Wu, H. Shen, C. Barugkin, F. J. Beck, T. P. White, K. Weber, and K. R. Catchpole, "Hysteresis phenomena in perovskite solar cells: the many and varied effects of ionic accumulation," Physical Chemistry Chemical Physics **19**, 3094–3103 (2017).

[217] P. Lopez-Varo, J. A. Jiménez-Tejada, M. García-Rosell, S. Ravishankar, G. Garcia-Belmonte, J. Bisquert, and O. Almora, "Device Physics of Hybrid Perovskite Solar cells: Theory and Experiment," Advanced Energy Materials **8**, 1702772 (2018).

[218] N. Marinova, S. Valero, and J. L. Delgado, "Organic and perovskite solar cells: Working principles, materials and interfaces," Journal of Colloid and Interface Science **488**, 373–389 (2017).

[219] G. Richardson, S. E. O'Kane, R. G. Niemann, T. A. Peltola, J. M. Foster, P. J. Cameron, and A. B. Walker, "Can slow-moving ions explain hysteresis in the current–voltage curves of perovskite solar cells?" Energy and Environmental Science **9**, 1476–1485 (2016).

[220] S. E. O'Kane, G. Richardson, A. Pockett, R. G. Niemann, J. M. Cave, N. Sakai, G. E. Eperon, H. J. Snaith, J. M. Foster, P. J. Cameron, and A. B. Walker, "Measurement and modelling of dark current decay transients in perovskite solar cells," Journal of Materials Chemistry C **5**, 452–462 (2017).

[221] P. Calado, A. M. Telford, D. Bryant, X. Li, J. Nelson, B. C. O'Regan, and P. R. Barnes, "Evidence for ion migration in hybrid perovskite solar cells with minimal hysteresis," Nature Communications **7**, 13831 (2016).

[222] Y. B. Martynov, R. G. Nazmitdinov, A. Moià-Pol, P. P. Gladyshev, A. R. Tameev, A. V. Vannikov, and M. Pudlak, "On the efficiency limit of $ZnO/CH_3NH_3PbI_3/CuI$ perovskite solar cells," Physical Chemistry Chemical Physics **19**, 19916–19921 (2017).

[223] T. Wahl, J. Hanisch, S. Meier, M. Schultes, and E. Ahlswede, "Sputtered indium zinc oxide rear electrodes for inverted semitransparent perovskite solar cells without using a protective buffer layer," Organic Electronics **54**, 48–53 (2018).

[224] L. Meng, J. You, T. F. Guo, and Y. Yang, "Recent Advances in the Inverted Planar Structure of Perovskite Solar Cells," Accounts of Chemical Research **49**, 155–165 (2016).

[225] J. You, L. Meng, Z. Hong, G. Li, and Y. Yang, "Inverted Planar Structure of Perovskite Solar Cells," in "Organic–Inorganic Halide Perovskite Photovoltaics: From Fundamentals to Device Architectures," , N. G. Park, M. Grätzel, and T. Miyasaka, eds. (Springer International Publishing, Switzerland, 2016), chap. 12, pp. 307–324.

[226] T. Liu, K. Chen, Q. Hu, R. Zhu, and Q. Gong, "Inverted Perovskite Solar Cells: Progresses and Perspectives," Advanced Energy Materials **6**, 1600457 (2016).

[227] G. Kakavelakis, T. Maksudov, D. Konios, I. Paradisanos, G. Kioseoglou, E. Stratakis, and E. Kymakis, "Efficient and Highly Air Stable Planar Inverted Perovskite Solar Cells with Reduced Graphene Oxide Doped PCBM Electron Transporting Layer," Advanced Energy Materials **7**, 1602120 (2017).

[228] W. Chen, F. Z. Liu, X. Y. Feng, A. B. Djurišić, W. K. Chan, and Z. B. He, "Cesium Doped NiO_x as an Efficient Hole Extraction Layer for Inverted Planar Perovskite Solar Cells," Advanced Energy Materials **7**, 1700722 (2017).

[229] J. H. Heo, H. J. Han, D. Kim, T. K. Ahn, and S. H. Im, "Hysteresis-less inverted $CH_3NH_3PbI_3$ planar perovskite hybrid solar cells with 18.1% power conversion efficiency," Energy and Environmental Science **8**, 1602–1608 (2015).

[230] K. Yao, F. Li, Q. He, X. Wang, Y. Jiang, H. Huang, and A. K. Jen, "A copper-doped nickel oxide bilayer for enhancing efficiency and stability of hysteresis-free inverted mesoporous perovskite solar cells," Nano Energy **40**, 155–162 (2017).

[231] L. Calió, S. Kazim, M. Grätzel, and S. Ahmad, "Hole-Transport Materials for Perovskite Solar Cells," Angewandte Chemie – International Edition **55**, 14522–14545 (2016).

[232] T. Abzieher, S. Moghadamzadeh, F. Schackmar, H. Eggers, F. Sutterlüti, A. Farooq, D. Kojda, K. Habicht, R. Schmager, A. Mertens, R. Azmi, L. Klohr, J. A. Schwenzer, M. Hetterich, U. Lemmer, B. S. Richards, M. Powalla, and U. W. Paetzold, "Electron-Beam-Evaporated Nickel Oxide Hole Transport Layers for Perovskite-Based Photovoltaics," Advanced Energy Materials **9**, 1802995 (2019).

[233] N. Arora, M. I. Dar, A. Hinderhofer, N. Pellet, F. Schreiber, S. M. Zakeeruddin, and M. Grätzel, "Perovskite solar cells with CuSCN hole extraction layers yield stabilized efficiencies greater than 20%," Science **358**, 768–771 (2017).

[234] Q. Wang, C. Bi, and J. Huang, "Doped hole transport layer for efficiency enhancement in planar heterojunction organolead trihalide perovskite solar cells," Nano Energy **15**, 275–280 (2015).

[235] X. Meng, Y. Bai, S. Xiao, T. Zhang, C. Hu, Y. Yang, X. Zheng, and S. Yang, "Designing new fullerene derivatives as electron transporting materials for efficient perovskite solar cells with improved moisture resistance," Nano Energy **30**, 341–346 (2016).

[236] M. L. Petrus, M. T. Sirtl, A. C. Closs, T. Bein, and P. Docampo, "Hydrazone-based hole transporting material prepared via condensation chemistry as alternative for cross-coupling chemistry for perovskite solar cells," Molecular Systems Design and Engineering **3**, 734–740 (2018).

[237] M. L. Petrus, K. Schutt, M. T. Sirtl, E. M. Hutter, A. C. Closs, J. M. Ball, J. C. Bijleveld, A. Petrozza, T. Bein, T. J. Dingemans, T. J. Savenije, H. Snaith, and P. Docampo, "New Generation Hole Transporting Materials for Perovskite Solar Cells: Amide-Based Small-Molecules with Nonconjugated Backbones," Advanced Energy Materials **8**, 1801605 (2018).

[238] N. Giesbrecht, J. Schlipf, L. Oesinghaus, A. Binek, T. Bein, P. Müller-Buschbaum, and P. Docampo, "Synthesis of Perfectly Oriented and Micrometer-Sized $MAPbBr_3$ Perovskite Crystals for Thin-Film Photovoltaic Applications," ACS Energy Letters **1**, 150–154 (2016).

[239] N. Giesbrecht, J. Schlipf, I. Grill, P. Rieder, V. Dyakonov, T. Bein, A. Hartschuh, P. Müller-Buschbaum, and P. Docampo, "Single-crystal-like optoelectronic-properties of $MAPbI_3$ perovskite polycrystalline thin films," Journal of Materials Chemistry A **6**, 4822–4828 (2018).

[240] M. Schultes, N. Giesbrecht, J. Küffner, E. Ahlswede, P. Docampo, T. Bein, and M. Powalla, "Universal Nanoparticle Wetting Agent for Upscaling Perovskite Solar Cells," ACS Applied Materials and Interfaces **11**, 12948–12957 (2019).

[241] F. Ruf, P. Rietz, M. F. Aygüler, I. Kelz, P. Docampo, H. Kalt, and M. Hetterich, "The Bandgap as a Moving Target: Reversible Bandgap Instabilities in Multiple-Cation Mixed-Halide Perovskite Solar Cells," ACS Energy Letters **3**, 2995–3001 (2018).

[242] C. Momblona, L. Gil-Escrig, E. Bandiello, E. M. Hutter, M. Sessolo, K. Lederer, J. Blochwitz-Nimoth, and H. J. Bolink, "Efficient vacuum deposited p–i–n and n–i–p perovskite solar cells employing doped charge transport layers," Energy and Environmental Science **9**, 3456–3463 (2016).

[243] X. Zhu, D. Yang, R. Yang, B. Yang, Z. Yang, X. Ren, J. Zhang, J. Niu, J. Feng, and S. Liu, "Superior stability for perovskite solar cells with 20% efficiency using vacuum co-evaporation," Nanoscale **9**, 12316–12323 (2017).

[244] W. Ke, D. Zhao, C. R. Grice, A. J. Cimaroli, G. Fang, and Y. Yan, "Efficient fully-vacuum-processed perovskite solar cells using copper phthalocyanine as hole selective layers," Journal of Materials Chemistry A **3**, 23888–23894 (2015).

[245] G. Longo, C. Momblona, M. G. La-Placa, L. Gil-Escrig, M. Sessolo, and H. J. Bolink, "Fully Vacuum-Processed Wide Band Gap Mixed-Halide Perovskite Solar Cells," ACS Energy Letters **3**, 214–219 (2018).

[246] L. A. Frolova, D. V. Anokhin, A. A. Piryazev, S. Y. Luchkin, N. N. Dremova, K. J. Stevenson, and P. A. Troshin, "Highly Efficient All-Inorganic Planar Heterojunction Perovskite Solar Cells Produced by Thermal Coevaporation of CsI and PbI_2," Journal of Physical Chemistry Letters **8**, 67–72 (2017).

[247] H. Ishihara, W. Chen, Y. C. Chen, S. Sarang, N. De Marco, O. Lin, S. Ghosh, and V. Tung, "Electrohydrodynamically Assisted Deposition of Efficient Perovskite Photovoltaics," Advanced Materials Interfaces **3**, 1500762 (2016).

[248] J. H. Heo, M. H. Lee, M. H. Jang, and S. H. Im, "Highly efficient $CH_3NH_3PbI_{3-x}Cl_x$ mixed halide perovskite solar cells prepared by re-dissolution and crystal grain growth via spray coating," Journal of Materials Chemistry A **4**, 17636–17642 (2016).

[249] K. Hwang, Y. S. Jung, Y. J. Heo, F. H. Scholes, S. E. Watkins, J. Subbiah, D. J. Jones, D. Y. Kim, and D. Vak, "Toward Large Scale Roll-to-Roll Production of Fully Printed Perovskite Solar Cells," Advanced Materials **27**, 1241–1247 (2015).

[250] T. M. Schmidt, T. T. Larsen-Olsen, J. E. Carlé, D. Angmo, and F. C. Krebs, "Upscaling of Perovskite Solar Cells: Fully Ambient Roll Processing of Flexible Perovskite Solar Cells with Printed Back Electrodes," Advanced Energy Materials **5**, 1500569 (2015).

[251] T. Qin, W. Huang, J. E. Kim, D. Vak, C. Forsyth, C. R. McNeill, and Y. B. Cheng, "Amorphous hole-transporting layer in slot-die coated perovskite solar cells," Nano Energy **31**, 210–217 (2017).

[252] S. Tang, Y. Deng, X. Zheng, Y. Bai, Y. Fang, Q. Dong, H. Wei, and J. Huang, "Composition Engineering in Doctor-Blading of Perovskite Solar Cells," Advanced Energy Materials **7**, 1700302 (2017).

[253] M. Yang, Z. Li, M. O. Reese, O. G. Reid, D. H. Kim, S. Siol, T. R. Klein, Y. Yan, J. J. Berry, M. F. A. M. van Hest, and K. Zhu, "Perovskite ink with wide processing window for scalable high-efficiency solar cells," Nature Energy **2**, 17038 (2017).

[254] S. Razza, F. Di Giacomo, F. Matteocci, L. Cinà, A. L. Palma, S. Casaluci, P. Cameron, A. D'Epifanio, S. Licoccia, A. Reale, T. M. Brown, and A. Di Carlo, "Perovskite solar cells and large area modules ($100\,cm^2$) based on an air flow-assisted PbI_2 blade coating deposition process," Journal of Power Sources **277**, 286–291 (2015).

[255] K. Cao, Z. Zuo, J. Cui, Y. Shen, T. Moehl, S. M. Zakeeruddin, M. Grätzel, and M. Wang, "Efficient screen printed perovskite solar cells based on mesoscopic TiO_2/Al_2O_3/NiO/carbon architecture," Nano Energy **17**, 171–179 (2015).

[256] F. Mathies, T. Abzieher, A. Hochstuhl, K. Glaser, A. Colsmann, U. W. Paetzold, G. Hernandez-Sosa, U. Lemmer, and A. Quintilla, "Multipass inkjet printed planar methylammonium lead iodide perovskite solar cells," Journal of Materials Chemistry A **4**, 19207–19213 (2016).

[257] F. Mathies, H. Eggers, B. S. Richards, G. Hernandez-Sosa, U. Lemmer, and U. W. Paetzold, "Inkjet-Printed Triple Cation Perovskite Solar Cells," ACS Applied Energy Materials **1**, 1834–1839 (2018).

[258] S. Bag, J. R. Deneault, and M. F. Durstock, "Aerosol-Jet-Assisted Thin-Film Growth of $CH_3NH_3PbI_3$ Perovskites – A Means to Achieve High Quality, Defect-Free Films for Efficient Solar Cells," Advanced Energy Materials **7**, 1701151 (2017).

[259] S. Schlisske, F. Mathies, D. Busko, N. Strobel, T. Rödlmeier, B. S. Richards, U. Lemmer, U. W. Paetzold, G. Hernandez-Sosa, and E. Klampaftis, "Design and Color Flexibility for Inkjet-Printed Perovskite Photovoltaics," ACS Applied Energy Materials **2**, 764–769 (2019).

[260] A. Cannavale, M. Hörantner, G. E. Eperon, H. J. Snaith, F. Fiorito, U. Ayr, and F. Martellotta, "Building integration of semitransparent perovskite-based solar cells: Energy performance and visual comfort assessment," Applied Energy **194**, 94–107 (2017).

[261] I. A. Howard, T. Abzieher, I. M. Hossain, H. Eggers, F. Schackmar, S. Ternes, B. S. Richards, U. Lemmer, and U. W. Paetzold, "Coated and Printed Perovskites for Photovoltaic Applications," Advanced Materials **31**, 1806702 (2019).

[262] H. J. Snaith, A. Abate, J. M. Ball, G. E. Eperon, T. Leijtens, N. K. Noel, S. D. Stranks, J. T. W. Wang, K. Wojciechowski, and W. Zhang, "Anomalous Hysteresis in Perovskite Solar Cells," Journal of Physical Chemistry Letters **5**, 1511–1515 (2014).

[263] E. L. Unger, E. T. Hoke, C. D. Bailie, W. H. Nguyen, A. R. Bowring, T. Heumüller, M. G. Christoforo, and M. D. McGehee, "Hysteresis and transient behavior in current–voltage measurements of hybrid-perovskite absorber solar cells," Energy and Environmental Science **7**, 3690–3698 (2014).

[264] A. K. Jena and T. Miyasaka, "Hysteresis Characteristics and Device Stability," in "Organic–Inorganic Halide Perovskite Photovoltaics: From Fundamentals to Device Architectures," , N. G. Park, M. Grätzel, and T. Miyasaka, eds. (Springer International Publishing, Switzerland, 2016), chap. 10, pp. 255–284.

[265] S. Ravishankar, O. Almora, C. Echeverría-Arrondo, E. Ghahremanirad, C. Aranda, A. Guerrero, F. Fabregat-Santiago, A. Zaban, G. Garcia-Belmonte, and J. Bisquert, "Surface Polarization Model for the Dynamic Hysteresis of Perovskite Solar Cells," Journal of Physical Chemistry Letters **8**, 915–921 (2017).

[266] W. Tress, J. P. Correa Baena, M. Saliba, A. Abate, and M. Graetzel, "Inverted Current–Voltage Hysteresis in Mixed Perovskite Solar Cells: Polarization, Energy Barriers, and Defect Recombination," Advanced Energy Materials **6**, 1600396 (2016).

[267] N. Weber, "Design and Commissioning of a Setup for Temperature-Dependent Electro-Optic Investigations of Perovskite Solar Cells," Bachelor's thesis, Karlsruhe Institute of Technology (2016).

[268] E. Zimmermann, K. K. Wong, M. Müller, H. Hu, P. Ehrenreich, M. Kohlstädt, U. Würfel, S. Mastroianni, G. Mathiazhagan, A. Hinsch, T. P. Gujar, M. Thelakkat, T. Pfadler, and L. Schmidt-Mende, "Characterization of perovskite solar cells: Towards a reliable measurement protocol," APL Materials **4**, 091901 (2016).

[269] A. J. Cimaroli, Y. Yu, C. Wang, W. Liao, L. Guan, C. R. Grice, D. Zhao, and Y. Yan, "Tracking the maximum power point of hysteretic perovskite solar cells using a predictive algorithm," Journal of Materials Chemistry C **5**, 10152–10157 (2017).

[270] L. Rakocevic, F. Ernst, N. T. Yimga, S. Vashishtha, T. Aernouts, T. Heumueller, C. J. Brabec, R. Gehlhaar, and J. Poortmans, "Reliable Performance Comparison of Perovskite Solar Cells Using Optimized Maximum Power Point Tracking," Solar RRL **3**, 1800287 (2019).

[271] M. Saliba, "Perovskite solar cells must come of age," Science **359**, 388–389 (2018).

[272] C. Li, A. Guerrero, Y. Zhong, A. Gräser, C. A. M. Luna, J. Köhler, J. Bisquert, R. Hildner, and S. Huettner, "Real-Time Observation of Iodide Ion Migration in Methylammonium Lead Halide Perovskites," Small **13**, 1701711 (2017).

[273] B. Chen, M. Yang, S. Priya, and K. Zhu, "Origin of J–V Hysteresis in Perovskite Solar Cells," Journal of Physical Chemistry Letters **7**, 905–917 (2016).

[274] W. Tress, N. Marinova, T. Moehl, S. M. Zakeeruddin, M. K. Nazeeruddin, and M. Grätzel, "Understanding the rate-dependent J–V hysteresis, slow time component, and aging in $CH_3NH_3PbI_3$ perovskite solar cells: the role of a compensated electric field," Energy and Environmental Science **8**, 995–1004 (2015).

[275] J. Wei, Y. Zhao, H. Li, G. Li, J. Pan, D. Xu, Q. Zhao, and D. Yu, "Hysteresis Analysis Based on the Ferroelectric Effect in Hybrid Perovskite Solar Cells," Journal of Physical Chemistry Letters **5**, 3937–3945 (2014).

[276] Y. Shao, Z. Xiao, C. Bi, Y. Yuan, and J. Huang, "Origin and elimination of photocurrent hysteresis by fullerene passivation in $CH_3NH_3PbI_3$ planar heterojunction solar cells," Nature Communications **5**, 5784 (2014).

[277] K. Wojciechowski, S. D. Stranks, A. Abate, G. Sadoughi, A. Sadhanala, N. Kopidakis, G. Rumbles, C. Z. Li, R. H. Friend, A. K. Y. Jen, and H. J. Snaith, "Heterojunction Modification for Highly Efficient Organic–Inorganic Perovskite Solar Cells," ACS Nano **8**, 12701–12709 (2014).

[278] S. A. Weber, I. M. Hermes, S. H. Turren-Cruz, C. Gort, V. W. Bergmann, L. Gilson, A. Hagfeldt, M. Graetzel, W. Tress, and R. Berger, "How the formation of interfacial charge causes hysteresis in perovskite solar cells," Energy and Environmental Science **11**, 2404–2413 (2018).

[279] R. T. Ginting, M. K. Jeon, K. J. Lee, W. Y. Jin, T. W. Kim, and J. W. Kang, "Degradation mechanism of planar-perovskite solar cells: correlating evolution of iodine distribution and photocurrent hysteresis," Journal of Materials Chemistry A **5**, 4527–4534 (2017).

[280] C. Wang, C. Xiao, Y. Yu, D. Zhao, R. A. Awni, C. R. Grice, K. Ghimire, I. Constantinou, W. Liao, A. J. Cimaroli, P. Liu, J. Chen, N. J. Podraza, C. S. Jiang, M. M. Al-Jassim, X. Zhao, and Y. Yan, "Understanding and Eliminating Hysteresis for Highly Efficient Planar Perovskite Solar Cells," Advanced Energy Materials **7**, 1700414 (2017).

[281] J. Cao, B. Wu, R. Chen, Y. Wu, Y. Hui, B.-W. Mao, and N. Zheng, "Efficient, Hysteresis-Free, and Stable Perovskite Solar Cells with ZnO as Electron-Transport Layer: Effect of Surface Passivation," Advanced Materials **30**, 1705596 (2018).

[282] M. F. Aygüler, A. G. Hufnagel, P. Rieder, M. Wussler, W. Jaegermann, T. Bein, V. Dyakonov, M. L. Petrus, A. Baumann, and P. Docampo, "Influence of Fermi Level Alignment with Tin Oxide on the Hysteresis of Perovskite Solar Cells," ACS Applied Materials and Interfaces **10**, 11414–11419 (2018).

[283] T. Kirchartz, "High open-circuit voltages in lead-halide perovskite solar cells: experiment, theory and open questions," Philosophical Transactions of the Royal Society A **377**, 20180286 (2019).

[284] M. Stolterfoht, C. M. Wolff, Y. Amir, A. Paulke, L. Perdigón-Toro, P. Caprioglio, and D. Neher, "Approaching the fill factor Shockley–Queisser limit in stable, dopant-free triple cation perovskite solar cells," Energy and Environmental Science **10**, 1530–1539 (2017).

[285] K. Wang, C. Liu, P. Du, J. Zheng, and X. Gong, "Bulk heterojunction perovskite hybrid solar cells with large fill factor," Energy and Environmental Science **8**, 1245–1255 (2015).

[286] N. J. Jeon, H. Na, E. H. Jung, T.-Y. Yang, Y. G. Lee, G. Kim, H.-W. Shin, S. Il Seok, J. Lee, and J. Seo, "A fluorene-terminated hole-transporting material for highly efficient and stable perovskite solar cells," Nature Energy **3**, 682–689 (2018).

[287] Z. Liu, L. Krückemeier, B. Krogmeier, B. Klingebiel, J. A. Márquez, S. Levcenko, S. Öz, S. Mathur, U. Rau, T. Unold, and T. Kirchartz, "Open-Circuit Voltages Exceeding 1.26 V in Planar Methylammonium Lead Iodide Perovskite Solar Cells," ACS Energy Letters **4**, 110–117 (2019).

[288] J.-F. Guillemoles, T. Kirchartz, D. Cahen, and U. Rau, "Guide for the perplexed to the Shockley–Queisser model for solar cells," Nature Photonics **13**, 501–505 (2019).

[289] T. J. Jacobsson, J. P. Correa-Baena, M. Pazoki, M. Saliba, K. Schenk, M. Grätzel, and A. Hagfeldt, "Exploration of the compositional space for mixed lead halogen perovskites for high efficiency solar cells," Energy and Environmental Science **9**, 1706–1724 (2016).

[290] Oxford PV, "Oxford PV perovskite solar cell achieves 28% efficiency," `https://www.oxfordpv.com/news/oxford-pv-perovskite-solar-cell-achieves-28-efficiency`, accessed 2019-06-15.

[291] M. Hutchins, "Tandem cells at 28%," `https://www.pv-magazine.com/2018/12/20/tandems-at-28/`, accessed 2019-06-15.

[292] P. Docampo, J. M. Ball, M. Darwich, G. E. Eperon, and H. J. Snaith, "Efficient organometal trihalide perovskite planar-heterojunction solar cells on flexible polymer substrates," Nature Communications **4**, 2761 (2013).

[293] D. Yang, R. Yang, J. Zhang, Z. Yang, S. Liu, and C. Li, "High efficiency flexible perovskite solar cells using superior low temperature TiO_2," Energy and Environmental Science **8**, 3208–3214 (2015).

[294] J. H. Heo, M. H. Lee, H. J. Han, B. R. Patil, J. S. Yu, and S. H. Im, "Highly efficient low temperature solution processable planar type $CH_3NH_3PbI_3$ perovskite flexible solar cells," Journal of Materials Chemistry A **4**, 1572–1578 (2016).

[295] G. E. Eperon, V. M. Burlakov, A. Goriely, and H. J. Snaith, "Neutral Color Semitransparent Microstructured Perovskite Solar Cells," ACS Nano **8**, 591–598 (2014).

[296] R. Li, X. Xiang, X. Tong, J. Zou, and Q. Li, "Wearable Double-Twisted Fibrous Perovskite Solar Cell," Advanced Materials **27**, 3831–3835 (2015).

[297] J. Deng, L. Qiu, X. Lu, Z. Yang, G. Guan, Z. Zhang, and H. Peng, "Elastic perovskite solar cells," Journal of Materials Chemistry A **3**, 21070–21076 (2015).

[298] M. Kaltenbrunner, G. Adam, E. D. Głowacki, M. Drack, R. Schwödiauer, L. Leonat, D. H. Apaydin, H. Groiss, M. C. Scharber, M. S. White, N. S. Sariciftci, and S. Bauer, "Flexible high power-per-weight perovskite solar cells with chromium oxide–metal contacts for improved stability in air," Nature Materials **14**, 1032–1039 (2015).

[299] F. Lang, N. H. Nickel, J. Bundesmann, S. Seidel, A. Denker, S. Albrecht, V. V. Brus, J. Rappich, B. Rech, G. Landi, and H. C. Neitzert, "Radiation Hardness and Self-Healing of Perovskite Solar Cells," Advanced Materials **28**, 8726–8731 (2016).

[300] Y. Miyazawa, M. Ikegami, H.-W. Chen, T. Ohshima, M. Imaizumi, K. Hirose, and T. Miyasaka, "Tolerance of Perovskite Solar Cell to High-Energy Particle Irradiations in Space Environment," iScience **2**, 148–155 (2018).

[301] E. Yablonovitch, O. D. Miller, and S. R. Kurtz, "A great solar cell also needs to be a great LED: External fluorescence leads to new efficiency record," AIP Conference Proceedings **1519**, 9–11 (2013).

[302] X. Dai, Z. Zhang, Y. Jin, Y. Niu, H. Cao, X. Liang, L. Chen, J. Wang, and X. Peng, "Solution-processed, high-performance light-emitting diodes based on quantum dots," Nature **515**, 96–99 (2014).

[303] D. Di, A. S. Romanov, L. Yang, J. M. Richter, J. P. H. Rivett, S. Jones, T. H. Thomas, M. A. Jalebi, R. H. Friend, M. Linnolahti, M. Bochmann, and D. Credgington, "High-performance light-emitting diodes based on carbene-metal-amides," Science **356**, 159–163 (2017).

[304] G.-Z. Lu, Z.-L. Tu, L. Liu, W.-W. Zhang, and Y.-X. Zheng, "Fast synthesis of iridium(III) complexes with sulfur containing ancillary ligand for high-performance green OLEDs with EQE over 31%," Journal of Materials Chemistry C **7**, 7273–7278 (2019).

[305] S. A. Veldhuis, T. C. Sum, M. Li, N. Mathews, P. P. Boix, S. G. Mhaisalkar, and N. Yantara, "Perovskite Materials for Light-Emitting Diodes and Lasers," Advanced Materials **28**, 6804–6834 (2016).

[306] J. Sun, J. Wu, X. Tong, F. Lin, Y. Wang, and Z. M. Wang, "Organic/Inorganic Metal Halide Perovskite Optoelectronic Devices beyond Solar Cells," Advanced Science **5**, 1700780 (2018).

[307] L. N. Quan, F. P. García de Arquer, R. P. Sabatini, and E. H. Sargent, "Perovskites for Light Emission," Advanced Materials **30**, 1801996 (2018).

[308] Q. Liao, X. Jin, and H. Fu, "Tunable Halide Perovskites for Miniaturized Solid-State Laser Applications," Advanced Optical Materials **7**, 1900099 (2019).

[309] A. Babayigit, A. Ethirajan, M. Muller, and B. Conings, "Toxicity of organometal halide perovskite solar cells," Nature Materials **15**, 247–251 (2016).

[310] I. R. Benmessaoud, A. L. Mahul-Mellier, E. Horváth, B. Maco, M. Spina, H. A. Lashuel, and L. Forró, "Health hazards of methylammonium lead iodide based perovskites: cytotoxicity studies," Toxicology Research **5**, 407–419 (2016).

[311] B. Hailegnaw, S. Kirmayer, E. Edri, G. Hodes, and D. Cahen, "Rain on methylammonium lead iodide based perovskites: Possible environmental effects of perovskite solar cells," Journal of Physical Chemistry Letters **6**, 1543–1547 (2015).

[312] Z. Xiao, Z. Song, and Y. Yan, "From Lead Halide Perovskites to Lead-Free Metal Halide Perovskites and Perovskite Derivatives," Advanced Materials **31**, 1803792 (2019).

[313] M. Lyu, J.-H. Yun, P. Chen, M. Hao, and L. Wang, "Addressing Toxicity of Lead: Progress and Applications of Low-Toxic Metal Halide Perovskites and Their Derivatives," Advanced Energy Materials **7**, 1602512 (2017).

[314] H. Hu, B. Dong, and W. Zhang, "Low-toxic metal halide perovskites: opportunities and future challenges," Journal of Materials Chemistry A **5**, 11436–11449 (2017).

[315] M. Konstantakou and T. Stergiopoulos, "A critical review on tin halide perovskite solar cells," Journal of Materials Chemistry A **5**, 11518–11549 (2017).

[316] Y. Liao, H. Liu, W. Zhou, D. Yang, Y. Shang, Z. Shi, B. Li, X. Jiang, L. Zhang, L. N. Quan, R. Quintero-Bermudez, B. R. Sutherland, Q. Mi, E. H. Sargent, and Z. Ning, "Highly Oriented Low-Dimensional Tin Halide Perovskites with Enhanced Stability and Photovoltaic Performance," Journal of the American Chemical Society **139**, 6693–6699 (2017).

[317] S. Shao, J. Liu, G. Portale, H.-H. Fang, G. R. Blake, G. H. ten Brink, L. J. A. Koster, and M. A. Loi, "Highly Reproducible Sn-Based Hybrid Perovskite Solar Cells with 9% Efficiency," Advanced Energy Materials **8**, 1702019 (2018).

[318] Z. Zhu, C.-C. Chueh, N. Li, C. Mao, and A. K.-Y. Jen, "Realizing Efficient Lead-Free Formamidinium Tin Triiodide Perovskite Solar Cells via a Sequential Deposition Route," Advanced Materials **30**, 1703800 (2018).

[319] F. Bai, Y. Hu, Y. Hu, T. Qiu, X. Miao, and S. Zhang, "Lead-free, air-stable ultrathin $Cs_3Bi_2I_9$ perovskite nanosheets for solar cells," Solar Energy Materials and Solar Cells **184**, 15–21 (2018).

[320] E. Greul, M. L. Petrus, A. Binek, P. Docampo, and T. Bein, "Highly stable, phase pure $Cs_2AgBiBr_6$ double perovskite thin films for optoelectronic applications," Journal of Materials Chemistry A **5**, 19972–19981 (2017).

[321] W. Gao, C. Ran, J. Xi, B. Jiao, W. Zhang, M. Wu, X. Hou, and Z. Wu, "High-Quality $Cs_2AgBiBr_6$ Double Perovskite Film for Lead-Free Inverted Planar Heterojunction Solar Cells with 2.2% Efficiency," ChemPhysChem **19**, 1696–1700 (2018).

[322] A. Binek, M. L. Petrus, N. Huber, H. Bristow, Y. Hu, T. Bein, and P. Docampo, "Recycling Perovskite Solar Cells to Avoid Lead Waste," ACS Applied Materials and Interfaces **8**, 12881–12886 (2016).

[323] J. M. Kadro, N. Pellet, F. Giordano, A. Ulianov, O. Müntener, J. Maier, M. Grätzel, and A. Hagfeldt, "Proof-of-concept for facile perovskite solar cell recycling," Energy and Environmental Science **9**, 3172–3179 (2016).

[324] J. A. Schwenzer, L. Rakocevic, R. Gehlhaar, T. Abzieher, S. Gharibzadeh, S. Moghadamzadeh, A. Quintilla, B. S. Richards, U. Lemmer, and U. W. Paetzold, "Temperature Variation-Induced Performance Decline of Perovskite Solar Cells," ACS Applied Materials and Interfaces **10**, 16390–16399 (2018).

[325] A. M. A. Leguy, Y. Hu, M. Campoy-Quiles, M. I. Alonso, O. J. Weber, P. Azarhoosh, M. van Schilfgaarde, M. T. Weller, T. Bein, J. Nelson, P. Docampo, and P. R. F. Barnes, "Reversible Hydration of $CH_3NH_3PbI_3$ in Films, Single Crystals, and Solar Cells," Chemistry of Materials **27**, 3397–3407 (2015).

[326] Y. Yang, M. Yang, Z. Li, R. Crisp, K. Zhu, and M. C. Beard, "Comparison of Recombination Dynamics in $CH_3NH_3PbBr_3$ and $CH_3NH_3PbI_3$ Perovskite Films: Influence of Exciton Binding Energy," Journal of Physical Chemistry Letters **6**, 4688–4692 (2015).

[327] J. A. Christians, P. A. Miranda Herrera, and P. V. Kamat, "Transformation of the Excited State and Photovoltaic Efficiency of $CH_3NH_3PbI_3$ Perovskite upon Controlled Exposure to Humidified Air," Journal of the American Chemical Society **137**, 1530–1538 (2015).

[328] N. Aristidou, I. Sanchez-Molina, T. Chotchuangchutchaval, M. Brown, L. Martinez, T. Rath, and S. A. Haque, "The Role of Oxygen in the Degradation of Methylammonium Lead Trihalide Perovskite Photoactive Layers," Angewandte Chemie – International Edition **54**, 8208–8212 (2015).

[329] D. Bryant, N. Aristidou, S. Pont, I. Sanchez-Molina, T. Chotchunangatchaval, S. Wheeler, J. R. Durrant, and S. A. Haque, "Light and oxygen induced degradation limits the operational stability of methylammonium lead triiodide perovskite solar cells," Energy and Environmental Science **9**, 1655–1660 (2016).

[330] N. Aristidou, C. Eames, I. Sanchez-Molina, X. Bu, J. Kosco, M. Saiful Islam, and S. A. Haque, "Fast oxygen diffusion and iodide defects mediate oxygen-induced degradation of perovskite solar cells," Nature Communications **8**, 15218 (2017).

[331] D. Li, G. Wang, H.-C. Cheng, C.-Y. Chen, H. Wu, Y. Liu, Y. Huang, and X. Duan, "Size-dependent phase transition in methylammonium lead iodide perovskite microplate crystals," Nature Communications **7**, 11330 (2016).

[332] C. D. Bailie, E. L. Unger, S. M. Zakeeruddin, M. Grätzel, and M. D. McGehee, "Melt-infiltration of spiro-OMeTAD and thermal instability of solid-state dye-sensitized solar cells," Physical Chemistry Chemical Physics **16**, 4864–4870 (2014).

[333] T. Malinauskas, D. Tomkute-Luksiene, R. Sens, M. Daskeviciene, R. Send, H. Wonneberger, V. Jankauskas, I. Bruder, and V. Getautis, "Enhancing Thermal Stability and Lifetime of Solid-State Dye-Sensitized Solar Cells via Molecular Engineering of the Hole-Transporting Material Spiro-OMeTAD," ACS Applied Materials and Interfaces **7**, 11107–11116 (2015).

[334] A. K. Jena, Y. Numata, M. Ikegami, and T. Miyasaka, "Role of spiro-OMeTAD in performance deterioration of perovskite solar cells at high temperature and reuse of the perovskite films to avoid Pb-waste," Journal of Materials Chemistry A **6**, 2219–2230 (2018).

[335] E. T. Hoke, D. J. Slotcavage, E. R. Dohner, A. R. Bowring, H. I. Karunadasa, and M. D. McGehee, "Reversible photo-induced trap formation in mixed-halide hybrid perovskites for photovoltaics," Chemical Science **6**, 613–617 (2015).

[336] W. Rehman, D. P. McMeekin, J. B. Patel, R. L. Milot, M. B. Johnston, H. J. Snaith, and L. M. Herz, "Photovoltaic mixed-cation lead mixed-halide perovskites: links between crystallinity, photo-stability and electronic properties," Energy and Environmental Science **10**, 361–369 (2017).

[337] K. A. Bush, K. Frohna, R. Prasanna, R. E. Beal, T. Leijtens, S. A. Swifter, and M. D. McGehee, "Compositional Engineering for Efficient Wide Band Gap Perovskites with Improved Stability to Photoinduced Phase Segregation," ACS Energy Letters **3**, 428–435 (2018).

[338] A. Miyata, A. Mitioglu, P. Plochocka, O. Portugall, J. T.-W. Wang, S. D. Stranks, H. J. Snaith, and R. J. Nicholas, "Direct measurement of the exciton binding energy and effective masses for charge carriers in organic–inorganic tri-halide perovskites," Nature Physics **11**, 582–587 (2015).

[339] M. Hirasawa, T. Ishihara, T. Goto, K. Uchida, and N. Miura, "Magnetoabsorption of the lowest exciton in perovskite-type compound $(CH_3NH_3)PbI_3$," Physica B **201**, 427–430 (1994).

[340] K. Tanaka, T. Takahashi, T. Ban, T. Kondo, K. Uchida, and N. Miura, "Comparative study on the excitons in lead-halide-based perovskite-type crystals $CH_3NH_3PbBr_3$ $CH_3NH_3PbI_3$," Solid State Communications **127**, 619–623 (2003).

[341] M. E. Ziffer, J. C. Mohammed, and D. S. Ginger, "Electroabsorption Spectroscopy Measurements of the Exciton Binding Energy, Electron–Hole Reduced Effective Mass, and Band Gap in the Perovskite $CH_3NH_3PbI_3$," ACS Photonics **3**, 1060–1068 (2016).

[342] M. Cardona, *Modulation Spectroscopy*, Solid State Physics, Supplement 11 (Academic Press, New York, NY, 1969).

[343] P. Y. Yu and M. Cardona, *Fundamentals of Semiconductors* (Springer, Berlin Heidelberg, 2001), 3rd ed.

[344] C. F. Klingshirn, *Semiconductor Optics* (Springer, Berlin Heidelberg, 2007), 3rd ed.

[345] H. Kuzmany, *Solid-State Spectroscopy* (Springer, Berlin Heidelberg, 1998).

[346] D. E. Aspnes, "Modulation Spectroscopy / Electric Field Effects on the Dielectric Function of Semiconductors," in "Optical Properties of Solids," , vol. 2 of *Handbook on Semiconductors*, M. Balkanski, ed. (North-Holland, Amsterdam, 1980).

[347] D. E. Aspnes, "Third-Derivative Modulation Spectroscopy with Low-Field Electroreflectance," Surface Science **37**, 418–442 (1973).

[348] K. Seeger, *Semiconductor Physics*, Springer Series in Solid-State Sciences Vol. 40 (Springer, Berlin Heidelberg, 1997), 5th ed.

[349] K. L. Shaklee and R. E. Nahory, "Valley–Orbit Splitting of Free Excitons? The Absorption Edge of Si," Physical Review Letters **24**, 942–945 (1970).

[350] H. Kalt, *Optical Properties of III–V Semiconductors*, Springer Series in Solid-State Sciences Vol. 120 (Springer, Berlin Heidelberg, 1996).

[351] M. Cadelano, M. Saba, N. Sestu, V. Sarritzu, D. Marongiu, F. Chen, R. Piras, F. Quochi, A. Mura, and G. Bongiovanni, "Photoexcitations and Emission Processes in Organometal Trihalide Perovskites," in "Perovskite Materials – Synthesis, Characterisation, Properties, and Applications," , L. Pan and G. Zhu, eds. (IntechOpen, 2016), chap. 12, pp. 377–399.

[352] N. Sestu, M. Cadelano, V. Sarritzu, F. Chen, D. Marongiu, R. Piras, M. Mainas, F. Quochi, M. Saba, A. Mura, and G. Bongiovanni, "Absorption F-Sum Rule for the Exciton Binding Energy in Methylammonium Lead Halide Perovskites," Journal of Physical Chemistry Letters **6**, 4566–4572 (2015).

[353] M. Sardela, ed., *Practical Materials Characterization* (Springer, New York, NY, 2014).

[354] T. J. C. Hosea, "Estimating Critical-Point Parameters of Modulated Reflectance Spectra," Physica Status Solidi (B) **189**, 531–542 (1995).

[355] F. H. Pollak, "Modulation Spectroscopy As A Technique For Semiconductor Characterization," Proceedings of SPIE **0276**, 142–156 (1981).

[356] F. H. Pollak, "Modulation Spectroscopy of Semiconductors and Semiconductor Microstructures," in "Optical Properties of Semiconductors," , vol. 2 of *Handbook on Semiconductors*, M. Balkanski, ed. (North-Holland, Amsterdam, 1994).

[357] O. J. Glembocki and B. V. Shanabrook, "Photoreflectance Spectroscopy of Microstructures," in "The Spectroscopy of Semiconductors," , vol. 36 of *Semiconductors and Semimetals*, R. K. Willardson, A. C. Beer, and E. R. Weber, eds. (Academic Press, Boston, 1992), chap. 4, pp. 222–292.

[358] F. Ruf, A. Magin, M. Schultes, M. F. Aygüler, P. Docampo, E. Ahlswede, H. Kalt, and M. Hetterich, "Temperature-Dependent Electromodulation Spectroscopy of Excitons in Perovskite Solar Cells," in "2018 IEEE 7th World Conference on Photovoltaic Energy Conversion (WCPEC)," (Waikoloa Village, HI, USA, 2018), pp. 1550–1554.

[359] W. Franz, "Einfluß eines elektrischen Feldes auf eine optische Absorptionskante," Zeitschrift für Naturforschung **13 a**, 484–489 (1958).

[360] L. V. Keldysh, "The Effect of a Strong Electric Field on the Optical Properties of Insulating Crystals," Journal of Experimental and Theoretical Physics (U.S.S.R.) **34**, 1138–1141 (1958).

[361] D. E. Aspnes, "Electric-Field Effects on Optical Absorption near Thresholds in Solids," Physical Review **147**, 554–566 (1966).

[362] K. Tharmalingam, "Optical Absorption in the Presence of a Uniform Field," Physical Review **130**, 2204–2206 (1963).

[363] D. E. Aspnes, P. Handler, and D. F. Blossey, "Interband Dielectric Properties of Solids in an Electric Field," Physical Review **166**, 921–933 (1968).

[364] F. H. Pollak and O. J. Glembocki, "Modulation Spectroscopy Of Semiconductor Microstructures: An Overview," Proceedings of SPIE **0946** (1988).

[365] H. Shen and F. H. Pollak, "Generalized Franz–Keldysh theory of electromodulation," Physical Review B **42**, 7097–7102 (1990).

[366] H. Shen and M. Dutta, "Franz–Keldysh oscillations in modulation spectroscopy," Journal of Applied Physics **78**, 2151–2176 (1995).

[367] R. A. Batchelor, A. C. Brown, and A. Hamnett, "Theoretical and experimental results for p-type GaAs electrolyte electroreflectance," Physical Review B **41**, 1401–1412 (1990).

[368] D. A. B. Miller, D. S. Chemla, and S. Schmitt-Rink, "Relation between electro-absorption in bulk semiconductors and in quantum wells: The quantum-confined Franz-Keldysh effect," Physical Review B **33**, 6976–6982 (1986).

[369] M. Grundmann, *The Physics of Semiconductors: An Introduction Including Nanophysics and Applications*, Graduate Texts in Physics (Springer, Cham, 2016), 3rd ed.

[370] D. E. Aspnes and J. E. Rowe, "Asymptotic Convolution Integral for Electric Field Effects on the Interband Dielectric Function," Solid State Communications **8**, 1145–1149 (1970).

[371] D. E. Aspnes and J. E. Rowe, "Resonant Nonlinear Optical Susceptibility: Electroreflectance in the Low-Field Limit," Physical Review B **5**, 4022–4030 (1972).

[372] G. U. Bublitz and S. G. Boxer, "Stark Spectroscopy: Applications in Chemistry, Biology, and Materials Science," Annual Review of Physical Chemistry **48**, 213–242 (1997).

[373] G. Lanzani, *The Photophysics behind Photovoltaics and Photonics* (Wiley-VCH, Weinheim, 2012).

[374] V. Roiati, E. Mosconi, A. Listorti, S. Colella, G. Gigli, and F. De Angelis, "Stark Effect in Perovskite/TiO_2 Solar Cells: Evidence of Local Interfacial Order," Nano Letters **14**, 2168–2174 (2014).

[375] M. Bouduban, A. Burgos-Caminal, and J. E. Moser, "Unveiling the Nature of Charge Carrier Interactions by Electroabsorption Spectroscopy: An Illustration with Lead-Halide Perovskites," Chimia **71**, 231–235 (2017).

[376] B. V. Shanabrook, O. J. Glembocki, and W. T. Beard, "Photoreflectance modulation mechanisms in GaAs–$Al_xGa_{1-x}As$ multiple quantum wells," Physical Review B **35**, 2540–2543 (1987).

[377] O. J. Glembocki and B. V. Shanabrook, "Electromodulation Spectroscopy of Confined Systems," Superlattices and Microstructures **5**, 603–607 (1989).

[378] D. F. Blossey, "Wannier Exciton in an Electric Field. II. Electroabsorption in Direct-Band-Gap Solids," Physical Review B **3**, 1382–1391 (1971).

[379] J. E. Rowe and D. E. Aspnes, "Approximate Treatment of Exciton Effects in Electric Field Modulation via the Slater–Koster Interaction," Physical Review Letters **25**, 162–165 (1970).

[380] J. Misiewicz, G. Sek, R. Kudrawiec, and P. Sitarek, "Photomodulated reflectance and transmittance: optical characterisation of novel semiconductor materials and device structures," Thin Solid Films **450**, 14–22 (2004).

[381] Y. S. Huang, H. Qiang, F. H. Pollak, J. Lee, and B. Elman, "Electroreflectance study of a symmetrically coupled GaAs/$Ga_{0.77}Al_{0.23}As$ double quantum well system," Journal of Applied Physics **70**, 3808–3814 (1991).

[382] A. W. Grau, *Bandstrukturuntersuchungen an Ga(In)NAs-basierten Halbleiterhetero-strukturen mittels Modulations- und magnetooptischer Spektroskopie* (Cuvillier Verlag, Göttingen, and Dissertation, Universität Karlsruhe, 2006).

[383] P. Rietz, "Characterization of Reversible Band Gap Shifts in Perovskite Solar Cells Investigated by Electroreflectance Spectroscopy," Bachelor's thesis, Karlsruhe Institute of Technology (2018).

[384] T. Yamamoto, T. Toyama, and H. Okamoto, "Electroreflectance Spectrum of $CdTe_{1-x}S_x$ Mixed Crystal Layer in CdS/CdTe Thin-Film Solar Cell," Japanese Journal of Applied Physics **37**, L916–L918 (1998).

[385] R. G. Dhere, Y. Zhang, M. J. Romero, M. Young, B. To, and R. Noufi, "Investigation of Junction Properties of CdS/CdTe Solar Cells and Their Correlation to Device Properties," in "2008 33rd IEEE Photovoltaic Specialists Conference," (San Diego, CA, USA, 2008).

[386] E. Cánovas, D. Fuertes Marrón, A. Martí, A. Luque, A. W. Bett, F. Dimroth, and S. P. Philipps, "Photoreflectance analysis of a GaInP/GaInAs/Ge multijunction solar cell," Applied Physics Letters **97**, 203504 (2010).

[387] J. Krustok, T. Raadik, M. Grossberg, S. Giraldo, M. Neuschitzer, S. López-Marino, and E. Saucedo, "Temperature dependent electroreflectance study of $Cu_2ZnSnSe_4$ solar cells," Materials Science in Semiconductor Processing **39**, 251–254 (2015).

[388] I. Kelz, "Electroreflectance Spectroscopy of Organic–Inorganic Halide Perovskite Solar Cells," Master's thesis, Karlsruhe Institute of Technology (2017).

[389] A. Magin, "Temperature-Dependent Electromodulation Spectroscopy of Organic–Inorganic Halide Perovskite Solar Cells," Master's thesis, Karlsruhe Institute of Technology (2018).

[390] M. van Eerden, M. Jaysankar, A. Hadipour, T. Merckx, J. J. Schermer, T. Aernouts, J. Poortmans, and U. W. Paetzold, "Optical Analysis of Planar Multicrystalline Perovskite Solar Cells," Advanced Optical Materials **5**, 1700151 (2017).

[391] N. Marinova, W. Tress, R. Humphry-Baker, M. I. Dar, V. Bojinov, S. M. Zakeeruddin, M. K. Nazeeruddin, and M. Grätzel, "Light Harvesting and Charge Recombination in $CH_3NH_3PbI_3$ Perovskite Solar Cells Studied by Hole Transport Layer Thickness Variation," ACS Nano **9**, 4200–4209 (2015).

[392] P. M. Kaminski, P. J. Isherwood, G. Womack, and J. M. Walls, "Optical Optimization of Perovskite Solar Cell Structure for Maximum Current Collection," Energy Procedia **102**, 11–18 (2016).

[393] X. Wu, H. Yu, N. Li, F. Wang, H. Xu, and N. Zhao, "Composition-Dependent Light-Induced Dipole Moment Change in Organometal Halide Perovskites," Journal of Physical Chemistry C **119**, 1253–1259 (2015).

[394] University of Gent, "Simulation programme SCAPS-1D for thin film solar cells," `http://scaps.elis.ugent.be/`, accessed 2019-04-30.

[395] M. Burgelman, P. Nollet, and S. Degrave, "Modelling polycrystalline semiconductor solar cells," Thin Solid Films **361–362**, 527–532 (2000).

[396] M. Burgelman, K. Decock, S. Kheli, and A. Abass, "Advanced electrical simulation of thin film solar cells," Thin Solid Films **535**, 296–301 (2013).

[397] C. Huber, "Electroreflectance Spectroscopy on Cu(In,Ga)Se_2 and $Cu_2ZnSn(S,Se)_4$ in Thin-Film Solar Cells: Experiment and Simulation," Master's thesis, Karlsruhe Institute of Technology (2015).

[398] C. Krämmer, *Optoelectronic Characterization of Thin-Film Solar Cells by Electroreflectance and Luminescence Spectroscopy* (Shaker Verlag, Düren, and Dissertation, Karlsruhe Institute of Technology, 2015).

[399] B. A. Gregg, "Excitonic Solar Cells," Journal of Physical Chemistry B **107**, 4688–4698 (2003).

[400] J. Tauc, R. Grigorovici, and A. Vancu, "Optical Properties and Electronic Structure of Amorphous Germanium," Physica Status Solidi **15**, 627–637 (1966).

[401] E. A. Davis and N. F. Mott, "Conduction in non-crystalline systems – V. Conductivity, optical absorption and photoconductivity in amorphous semiconductors," Philosophical Magazine **22**, 903–922 (1970).

[402] B. D. Viezbicke, S. Patel, B. E. Davis, and D. P. Birnie, "Evaluation of the Tauc method for optical absorption edge determination: ZnO thin films as a model system," Physica Status Solidi (B) **252**, 1700–1710 (2015).

[403] A. M. Askar and K. Shankar, "Exciton Binding Energy in Organic-Inorganic Tri-Halide Perovskites," Journal of Nanoscience and Nanotechnology **16**, 5890–5901 (2016).

[404] X. Zhu, H. Su, R. A. Marcus, and M. E. Michel-Beyerle, "Computed and Experimental Absorption Spectra of the Perovskite $CH_3NH_3PbI_3$," Journal of Physical Chemistry Letters **5**, 3061–3065 (2014).

[405] U. G. Jong, C. J. Yu, Y. M. Jang, G. C. Ri, S. N. Hong, and Y. H. Pae, "Revealing the stability and efficiency enhancement in mixed halide perovskites $MAPb(I_{1-x}Cl_x)_3$ with ab initio calculations," Journal of Power Sources **350**, 65–72 (2017).

[406] Q. Lin, A. Armin, R. C. R. Nagiri, P. L. Burn, and P. Meredith, "Electro-optics of perovskite solar cells," Nature Photonics **9**, 106–112 (2014).

[407] K. Galkowski, A. Mitioglu, A. Miyata, P. Plochocka, O. Portugall, G. E. Eperon, J. T.-W. Wang, T. Stergiopoulos, S. D. Stranks, H. J. Snaith, and R. J. Nicholas, "Determination of the exciton binding energy and effective masses for methylammonium and formamidinium lead tri-halide perovskite semiconductors," Energy and Environmental Science **9**, 962–970 (2016).

[408] Z. Yang, A. Surrente, K. Galkowski, N. Bruyant, D. K. Maude, A. A. Haghighirad, H. J. Snaith, P. Plochocka, and R. J. Nicholas, "Unraveling the Exciton Binding Energy and the Dielectric Constant in Single-Crystal Methylammonium Lead Triiodide Perovskite," Journal of Physical Chemistry Letters **8**, 1851–1855 (2017).

[409] S. Sun, T. Salim, N. Mathews, M. Duchamp, C. Boothroyd, G. Xing, T. C. Sum, and Y. M. Lam, "The origin of high efficiency in low-temperature solution-processable bilayer organometal halide hybrid solar cells," Energy and Environmental Science **7**, 399–407 (2014).

[410] T. J. Savenije, C. S. Ponseca, L. Kunneman, M. Abdellah, K. Zheng, Y. Tian, Q. Zhu, S. E. Canton, I. G. Scheblykin, T. Pullerits, A. Yartsev, and V. Sundström, "Thermally Activated Exciton Dissociation and Recombination Control the Carrier Dynamics in Organometal Halide Perovskite," Journal of Physical Chemistry Letters **5**, 2189–2194 (2014).

[411] T. Ishihara, "Optical properties of PbI-based perovskite structures," Journal of Luminescence **61–62**, 269–274 (1994).

[412] V. D'Innocenzo, G. Grancini, M. J. P. Alcocer, A. R. S. Kandada, S. D. Stranks, M. M. Lee, G. Lanzani, H. J. Snaith, and A. Petrozza, "Excitons versus free charges in organo-lead tri-halide perovskites," Nature Communications **5**, 3586 (2014).

[413] D. J. Slotcavage, H. I. Karunadasa, and M. D. McGehee, "Light-Induced Phase Segregation in Halide-Perovskite Absorbers," ACS Energy Letters **1**, 1199–1205 (2016).

[414] A. Sadhanala, F. Deschler, T. H. Thomas, S. E. Dutton, K. C. Goedel, F. C. Hanusch, M. L. Lai, U. Steiner, T. Bein, P. Docampo, D. Cahen, and R. H. Friend, "Preparation of Single-Phase Films of $CH_3NH_3Pb(I_{1-x}Br_x)_3$ with Sharp Optical Band Edges," Journal of Physical Chemistry Letters **5**, 2501–2505 (2014).

[415] C. Sheng, K. Mielczarek, W. Ma, W. Wang, Z. V. Vardeny, Y. Zhai, C. Zhang, and A. Zakhidov, "Exciton versus Free Carrier Photogeneration in Organometal Trihalide Perovskites Probed by Broadband Ultrafast Polarization Memory Dynamics," Physical Review Letters **114**, 116601 (2015).

[416] C. G. Wu, C. H. Chiang, and S. H. Chang, "A perovskite cell with a record-high-V_{OC} of 1.61 V based on solvent annealed $CH_3NH_3PbBr_3$/ICBA active layer," Nanoscale **8**, 4077–4085 (2016).

[417] S. S. Shin, E. J. Yeom, W. S. Yang, S. Hur, M. G. Kim, J. Im, J. Seo, J. H. Noh, and S. I. Seok, "Colloidally prepared La-doped $BaSnO_3$ electrodes for efficient, photostable perovskite solar cells," Science **356**, 167–171 (2017).

[418] A. Shah, P. Torres, R. Tscharner, N. Wyrsch, and H. Keppner, "Photovoltaic Technology: The Case for Thin-Film Solar Cells," Science **285**, 692–699 (1999).

[419] T. J. Coutts, K. A. Emery, and J. S. Ward, "Modeled Performance of Polycrystalline Thin-Film Tandem Solar Cells," Progress in Photovoltaics: Research and Applications **10**, 195–203 (2002).

[420] M. Filipic, P. Löper, B. Niesen, S. De Wolf, J. Krc, C. Ballif, and M. Topic, "$CH_3NH_3PbI_3$ perovskite / silicon tandem solar cells: characterization based optical simulations," Optics Express **23**, 263–278 (2015).

[421] M. H. Futscher and B. Ehrler, "Efficiency Limit of Perovskite/Si Tandem Solar Cells," ACS Energy Letters **1**, 863–868 (2016).

[422] P. Lange, "Temperature-Dependent Spectroscopy on Optoelectronical Properties of Perovskite Solar Cells," Master's thesis, Karlsruhe Institute of Technology (2019).

[423] D. Bahro, "Externe Quanteneffizienzmessungen an organischen Tandemsolarzellen," Dissertation, Karlsruhe Institute of Technology (2017).

[424] K. Mertens, *Photovoltaics: Fundamentals, Technology, and Practice* (Wiley, Chichester; Hoboken, NJ, 2019), 2nd ed.

[425] O. Malinkiewicz, A. Yella, Y. H. Lee, G. M. Espallargas, M. Graetzel, H. J. Bolink, and M. K. Nazeeruddin, "Perovskite solar cells employing organic charge-transport layers," Nature Photonics **8**, 128–132 (2013).

[426] J. B. Patel, Q. Lin, O. Zadvorna, C. L. Davies, L. M. Herz, and M. B. Johnston, "Photocurrent Spectroscopy of Perovskite Solar Cells over a Wide Temperature Range from 15 to 350 K," Journal of Physical Chemistry Letters **9**, 263–268 (2018).

[427] H. Zhang, X. Qiao, Y. Shen, T. Moehl, S. M. Zakeeruddin, M. Grätzel, and M. Wang, "Photovoltaic behaviour of lead methylammonium triiodide perovskite solar cells down to 80 K," Journal of Materials Chemistry A **3**, 11762–11767 (2015).

[428] P. Piatkowski, B. Cohen, C. S. Ponseca, M. Salado, S. Kazim, S. Ahmad, V. Sundström, and A. Douhal, "Unraveling Charge Carriers Generation, Diffusion, and Recombination in Formamidinium Lead Triiodide Perovskite Polycrystalline Thin Film," Journal of Physical Chemistry Letters **7**, 204–210 (2016).

[429] S. D. Stranks, V. M. Burlakov, T. Leijtens, J. M. Ball, A. Goriely, and H. J. Snaith, "Recombination Kinetics in Organic–Inorganic Perovskites: Excitons, Free Charge, and Subgap States," Physical Review Applied **2**, 034007 (2014).

[430] M. N. Saha, "LIII. Ionization in the solar chromosphere," The London, Edinburgh, and Dublin Philosophical Magazine and Journal of Science **40**, 472–488 (1920).

[431] M. N. Saha, "On a Physical Theory of Stellar Spectra," Proceedings of the Royal Society of London A **99**, 135–153 (1921).

[432] K. H. Kingdon and I. Langmuir, "The Removal of Thorium From the Surface of a Thoriated Tungsten Filament by Positive Ion Bombardment," Physical Review **22**, 148–160 (1923).

[433] M. J. Dresser, "The Saha-Langmuir Equation and its Application," Journal of Applied Physics **39**, 338–339 (1968).

[434] M. Combescot, "Thermodynamics of an Electron–Hole System in Semiconductors," Physica Status Solidi (B) **86**, 349–358 (1978).

[435] R. Cingolani, L. Calcagnile, G. Colí, R. Rinaldi, M. Lomoscolo, M. DiDio, A. Franciosi, L. Vanzetti, G. C. LaRocca, and D. Campi, "Radiative recombination processes in wide-band-gap II–VI quantum wells: the interplay between excitons and free carriers," Journal of thc Optical Society of America B **13**, 1268–1277 (1996).

[436] G. Grancini, A. R. Srimath Kandada, J. M. Frost, A. J. Barker, M. De Bastiani, M. Gandini, S. Marras, G. Lanzani, A. Walsh, and A. Petrozza, "Role of microstructure in the electron–hole interaction of hybrid lead halide perovskites," Nature Photonics **9**, 695–701 (2015).

[437] P. Jenkins, "Space PV Systems and Flight Demonstrations," in "Photovoltaic Solar Energy: From Fundamentals to Applications," , A. Reinders, P. Verlinden, W. van Sark, and A. Freundlich, eds. (John Wiley & Sons, Ltd, Chichester, West Sussex, 2017), chap. 9.2, pp. 444–454, 1st ed.

[438] Y. P. Varshni, "Temperature Dependence of the Energy Gap in Semiconductors," Physica **34**, 149–154 (1967).

[439] H. Y. Fan, "Temperature Dependence of the Energy Gap in Semiconductors," Physical Review **82**, 900–905 (1951).

[440] M. B. Panish and H. C. Casey, "Temperature Dependence of the Energy Gap in GaAs and GaP," Journal of Applied Physics **40**, 163–167 (1969).

[441] J. Camassel and D. Auvergne, "Temperature dependence of the fundamental edge of germanium and zinc-blende-type semiconductors," Physical Review B **12**, 3258–3267 (1975).

[442] K. P. O'Donnell and X. Chen, "Temperature dependence of semiconductor band gaps," Applied Physics Letters **58**, 2924–2926 (1991).

[443] M. I. Dar, G. Jacopin, S. Meloni, A. Mattoni, N. Arora, A. Boziki, S. M. Zakeeruddin, U. Rothlisberger, and M. Grätzel, "Origin of unusual bandgap shift and dual emission in organic-inorganic lead halide perovskites," Science Advances **2**, e1601156 (2016).

[444] B. J. Foley, D. L. Marlowe, K. Sun, W. A. Saidi, L. Scudiero, M. C. Gupta, and J. J. Choi, "Temperature dependent energy levels of methylammonium lead iodide perovskite," Applied Physics Letters **106**, 243904 (2015).

[445] C. Quarti, E. Mosconi, J. M. Ball, V. D'Innocenzo, C. Tao, S. Pathak, H. J. Snaith, A. Petrozza, and F. De Angelis, "Structural and optical properties of methylammonium lead iodide across the tetragonal to cubic phase transition: implications for perovskite solar cells," Energy and Environmental Science **9**, 155–163 (2016).

[446] W. Kong, Z. Ye, Z. Qi, B. Zhang, M. Wang, A. Rahimi-Iman, and H. Wu, "Characterization of an abnormal photoluminescence behavior upon crystal-phase transition of perovskite $CH_3NH_3PbI_3$," Physical Chemistry Chemical Physics **17**, 16405–16411 (2015).

[447] T. Baikie, Y. Fang, J. M. Kadro, M. Schreyer, F. Wei, S. G. Mhaisalkar, M. Graetzel, and T. J. White, "Synthesis and crystal chemistry of the hybrid perovskite $(CH_3NH_3)PbI_3$ for solid-state sensitised solar cell applications," Journal of Materials Chemistry A **1**, 5628–5641 (2013).

[448] F. Panzer, S. Baderschneider, T. P. Gujar, T. Unger, S. Bagnich, M. Jakoby, H. Bässler, S. Hüttner, J. Köhler, R. Moos, M. Thelakkat, R. Hildner, and A. Köhler, "Reversible Laser Induced Amplified Spontaneous Emission from Coexisting Tetragonal and Orthorhombic Phases in Hybrid Lead Halide Perovskites," Advanced Optical Materials **4**, 917–928 (2016).

[449] C. Wehrenfennig, M. Liu, H. J. Snaith, M. B. Johnston, and L. M. Herz, "Charge carrier recombination channels in the low-temperature phase of organic-inorganic lead halide perovskite thin films," APL Materials **2**, 081513 (2014).

[450] F. Chen, C. Zhu, C. Xu, P. Fan, F. Qin, A. Gowri Manohari, J. Lu, Z. Shi, Q. Xu, and A. Pan, "Crystal structure and electron transition underlying photoluminescence of methylammonium lead bromide perovskites," Journal of Materials Chemistry C **5**, 7739–7745 (2017).

[451] I. P. Swainson, R. P. Hammond, C. Soullière, O. Knop, and W. Massa, "Phase transitions in the perovskite methylammonium lead bromide, $CH_3ND_3PbBr_3$," Journal of Solid State Chemistry **176**, 97–104 (2003).

[452] L. Cojocaru, S. Uchida, Y. Sanehira, V. Gonzalez-Pedro, J. Bisquert, J. Nakazaki, T. Kubo, and H. Segawa, "Temperature Effects on the Photovoltaic Performance of Planar Structure Perovskite Solar Cells," Chemistry Letters **44**, 1557–1559 (2015).

[453] H. T. Yi, X. Wu, X. Zhu, and V. Podzorov, "Intrinsic Charge Transport across Phase Transitions in Hybrid Organo-Inorganic Perovskites," Advanced Materials **28**, 6509–6514 (2016).

[454] M. N. F. Hoque, N. Islam, Z. Li, G. Ren, K. Zhu, and Z. Fan, "Ionic and Optical Properties of Methylammonium Lead Iodide Perovskite across the Tetragonal–Cubic Structural Phase Transition," ChemSusChem **9**, 2692–2698 (2016).

[455] G. M. Bernard, R. E. Wasylishen, C. I. Ratcliffe, V. Terskikh, Q. Wu, J. M. Buriak, and T. Hauger, "Methylammonium Cation Dynamics in Methylammonium Lead Halide Perovskites: A Solid-State NMR Perspective," Journal of Physical Chemistry A **122**, 1560–1573 (2018).

[456] T. J. Jacobsson, L. J. Schwan, M. Ottosson, A. Hagfeldt, and T. Edvinsson, "Determination of Thermal Expansion Coefficients and Locating the Temperature-Induced Phase Transition in Methylammonium Lead Perovskites Using X-ray Diffraction," Inorganic Chemistry **54**, 10678–10685 (2015).

[457] R. Prasanna, A. Gold-Parker, T. Leijtens, B. Conings, A. Babayigit, H. G. Boyen, M. F. Toney, and M. D. McGehee, "Band Gap Tuning via Lattice Contraction and Octahedral Tilting in Perovskite Materials for Photovoltaics," Journal of the American Chemical Society **139**, 11117–11124 (2017).

[458] T. Oku, "Crystal Structures of $CH_3NH_3PbI_3$ and Related Perovskite Compounds Used for Solar Cells," in "Solar Cells – New Approaches and Reviews," , L. A. Kosyachenko, ed. (IntechOpen, 2015), chap. 3, pp. 77–101.

[459] K. Page, J. E. Siewenie, P. Quadrelli, and L. Malavasi, "Short-Range Order of Methylammonium and Persistence of Distortion at the Local Scale in $MAPbBr_3$ Hybrid Perovskite," Angewandte Chemie – International Edition **55**, 14320–14324 (2016).

[460] I. Cardinaletti, T. Vangerven, S. Nagels, R. Cornelissen, D. Schreurs, J. Hruby, J. Vodnik, D. Devisscher, J. Kesters, J. D'Haen, A. Franquet, V. Spampinato, T. Conard, W. Maes, W. Deferme, and J. V. Manca, "Organic and perovskite solar cells for space applications," Solar Energy Materials and Solar Cells **182**, 121–127 (2018).

[461] R. J. Sutton, M. R. Filip, A. A. Haghighirad, N. Sakai, B. Wenger, F. Giustino, and H. J. Snaith, "Cubic or Orthorhombic? Revealing the Crystal Structure of Metastable Black-Phase $CsPbI_3$ by Theory and Experiment," ACS Energy Letters **3**, 1787–1794 (2018).

[462] J. Y. Yang and M. Hu, "Temperature-Induced Large Broadening and Blue Shift in the Electronic Band Structure and Optical Absorption of Methylammonium Lead Iodide Perovskite," Journal of Physical Chemistry Letters **8**, 3720–3725 (2017).

[463] C. K. Møller, "The Structure of Perovskite-Like Cæsium Plumbo Trihalides," Matematisk-fysiske meddelelser Kongelige Danske Videnskabernes Selskab **32** (1959).

[464] S. J. Yoon, M. Kuno, and P. V. Kamat, "Shift Happens. How Halide Ion Defects Influence Photoinduced Segregation in Mixed Halide Perovskites," ACS Energy Letters **2**, 1507–1514 (2017).

[465] A. J. Barker, A. Sadhanala, F. Deschler, M. Gandini, S. P. Senanayak, P. M. Pearce, E. Mosconi, A. J. Pearson, Y. Wu, A. R. Srimath Kandada, T. Leijtens, F. De Angelis, S. E. Dutton, A. Petrozza, and R. H. Friend, "Defect-Assisted Photoinduced Halide Segregation in Mixed-Halide Perovskite Thin Films," ACS Energy Letters **2**, 1416–1424 (2017).

[466] M. C. Brennan, S. Draguta, P. V. Kamat, and M. Kuno, "Light-Induced Anion Phase Segregation in Mixed Halide Perovskites," ACS Energy Letters **3**, 204–213 (2018).

[467] X. Tang, M. Van Den Berg, E. Gu, A. Horneber, G. J. Matt, A. Osvet, A. J. Meixner, D. Zhang, and C. J. Brabec, "Local Observation of Phase Segregation in Mixed-Halide Perovskite," Nano Letters **18**, 2172–2178 (2018).

[468] W. Li, M. U. Rothmann, A. Liu, Z. Wang, Y. Zhang, A. R. Pascoe, J. Lu, L. Jiang, Y. Chen, F. Huang, Y. Peng, Q. Bao, J. Etheridge, U. Bach, and Y. B. Cheng, "Phase Segregation Enhanced Ion Movement in Efficient Inorganic $CsPbIBr_2$ Solar Cells," Advanced Energy Materials **7**, 1700946 (2017).

[469] J. H. Noh, S. H. Im, J. H. Heo, T. N. Mandal, and S. I. Seok, "Chemical Management for Colorful, Efficient, and Stable Inorganic–Organic Hybrid Nanostructured Solar Cells," Nano Letters **13**, 1764–1769 (2013).

[470] G. F. Samu, C. Janáky, and P. V. Kamat, "A Victim of Halide Ion Segregation. How Light Soaking Affects Solar Cell Performance of Mixed Halide Lead Perovskites," ACS Energy Letters **2**, 1860–1861 (2017).

[471] R. G. Balakrishna, S. M. Kobosko, and P. V. Kamat, "Mixed Halide Perovskite Solar Cells. Consequence of Iodide Treatment on Phase Segregation Recovery," ACS Energy Letters **3**, 2267–2272 (2018).

[472] T. Duong, H. K. Mulmudi, Y. Wu, X. Fu, H. Shen, J. Peng, N. Wu, H. T. Nguyen, D. Macdonald, M. Lockrey, T. P. White, K. Weber, and K. Catchpole, "Light and Electrically Induced Phase Segregation and Its Impact on the Stability of Quadruple Cation High Bandgap Perovskite Solar Cells," ACS Applied Materials and Interfaces **9**, 26859–26866 (2017).

[473] F. Brivio, C. Caetano, and A. Walsh, "Thermodynamic Origin of Photoinstability in the $CH_3NH_3Pb(I_{1-x}Br_x)_3$ Hybrid Halide Perovskite Alloy," Journal of Physical Chemistry Letters **7**, 1083–1087 (2016).

[474] C. G. Bischak, C. L. Hetherington, H. Wu, S. Aloni, D. F. Ogletree, D. T. Limmer, and N. S. Ginsberg, "Origin of Reversible Photoinduced Phase Separation in Hybrid Perovskites," Nano Letters **17**, 1028–1033 (2017).

[475] S. Draguta, O. Sharia, S. J. Yoon, M. C. Brennan, Y. V. Morozov, J. M. Manser, P. V. Kamat, W. F. Schneider, and M. Kuno, "Rationalizing the light-induced phase separation of mixed halide organic–inorganic perovskites," Nature Communications **8**, 200 (2017).

[476] A. F. Gualdrón-Reyes, S. J. Yoon, E. M. Barea, S. Agouram, V. Muñoz-Sanjosé, Á. M. Meléndez, M. E. Niño-Gómez, and I. Mora-Seró, "Controlling the Phase Segregation in Mixed Halide Perovskites through Nanocrystal Size," ACS Energy Letters **4**, 54–62 (2019).

[477] R. A. Belisle, K. A. Bush, L. Bertoluzzi, A. Gold-Parker, M. F. Toney, and M. D. McGehee, "Impact of Surfaces on Photoinduced Halide Segregation in Mixed-Halide Perovskites," ACS Energy Letters **3**, 2694–2700 (2018).

[478] R. E. Beal, D. J. Slotcavage, T. Leijtens, A. R. Bowring, R. A. Belisle, W. H. Nguyen, G. F. Burkhard, E. T. Hoke, and M. D. McGehee, "Cesium Lead Halide Perovskites with Improved Stability for Tandem Solar Cells," Journal of Physical Chemistry Letters **7**, 746–751 (2016).

[479] E. Wirth, F. Ruf, M. F. Aygüler, J. Hanisch, P. Docampo, E. Ahlswede, H. Kalt, and M. Hetterich, "Reversible Changes of the Bandgap Energy in Multiple-Cation Mixed-Halide Perovskite Solar Cells Under Illumination and Bias Investigated by Optical Spectroscopy," presented at "DPG Spring Meeting," (Regensburg, Germany, 2019).

[480] J. M. Howard, E. M. Tennyson, S. Barik, R. Szostak, E. Waks, M. F. Toney, A. F. Nogueira, B. R. Neves, and M. S. Leite, "Humidity-Induced Photoluminescence Hysteresis in Variable Cs/Br Ratio Hybrid Perovskites," Journal of Physical Chemistry Letters **9**, 3463–3469 (2018).

[481] U. G. Jong, C. J. Yu, G. C. Ri, A. P. McMahon, N. M. Harrison, P. R. Barnes, and A. Walsh, "Influence of water intercalation and hydration on chemical decomposition and ion transport in methylammonium lead halide perovskites," Journal of Materials Chemistry A **6**, 1067–1074 (2018).

[482] R. Ciesielski, F. Schäfer, N. F. Hartmann, N. Giesbrecht, T. Bein, P. Docampo, and A. Hartschuh, "Grain Boundaries Act as Solid Walls for Charge Carrier Diffusion in Large Crystal MAPI Thin Films," ACS Applied Materials and Interfaces **10**, 7974–7981 (2018).

[483] A. Biewald, N. Giesbrecht, T. Bein, P. Docampo, A. Hartschuh, and R. Ciesielski, "Temperature-Dependent Ambipolar Charge Carrier Mobility in Large-Crystal Hybrid Halide Perovskite Thin Films," ACS Applied Materials and Interfaces **11**, 20838–20844 (2019).

[484] J. Liang, Z. Liu, L. Qiu, Z. Hawash, L. Meng, Z. Wu, Y. Jiang, L. K. Ono, and Y. Qi, "Enhancing Optical, Electronic, Crystalline, and Morphological Properties of Cesium Lead Halide by Mn Substitution for High-Stability All-Inorganic Perovskite Solar Cells," Advanced Energy Materials **8**, 1800504 (2018).

[485] C. Dong, X. Han, W. Li, Q. Qiu, and J. Wang, "Anti-solvent assisted multi-step deposition for efficient and stable carbon-based $CsPbI_2Br$ all-inorganic perovskite solar cell," Nano Energy **59**, 553–559 (2019).

[486] E. Förstner, A. Magin, F. Ruf, C. Huber, W. Witte, D. Hariskos, O. Kiowski, H. Kalt, and M. Hetterich, "Simulation of Electroreflectance Spectra of CIGS Solar Cells," presented at "DPG Spring Meeting," (Regensburg, Germany, 2019).

[487] M. C. Troparevsky, A. S. Sabau, A. R. Lupini, and Z. Zhang, "Transfer-matrix formalism for the calculation of optical response in multilayer systems: from coherent to incoherent interference," Optics Express **18**, 24715 (2010).

[488] W. H. Bragg and W. L. Bragg, "The Reflection of X-rays by Crystals," Proceedings of the Royal Society of London A **88**, 428–438 (1913).

[489] A. Benninghoven, "Chemical Analysis of Inorganic and Organic Surfaces and Thin Films by Static Time-of-Flight Secondary Ion Mass Spectrometry (TOF-SIMS)," Angewandte Chemie – International Edition **33**, 1023–1043 (1994).

Closing Words

Without any doubt, this work would not have been possible without the support of various people – both scientifically and beyond. That's why I want to use these final pages to say thank you:

- **Prof. Dr. Heinz Kalt** and **Priv.-Doz. Dr. Michael Hetterich** for the privilege to be part of this group and to work on this interesting research topic. Thank you for your patience and time, for numerous valuable discussions, for the freedom to help shape this research project according to my ideas, and for many opportunities to present our results at different conferences.
- "My" bachelor and master students **Nico Weber**, **Ina Kelz**, **Alice Magin**, **Pascal Rietz**, **Philip Lange**, **Eva Wirth**, and **Ellen Förstner**. Thank you for all your work, your dedication and inspiring input for the perovskite research and everything we shared beyond that. I definitely learned a lot together with and from you.
- My colleagues and friends **Tobias Siegle** and **Simon Woska** for the amazing time we shared in our office and all you have invested in our group. It was a pleasure to work with you and to share memorable moments at university and beyond.
- **Sarah Krämmer**, **Christoph Krämmer**, and **Mario Lang** for all your helpful advice and everything you invested in our group on which I and many others could build on.
- **All current and former members of our working group** for the friendly atmosphere – inside and outside of university. I am especially grateful for many meaningful personal conversations beyond scientific topics thanks to **Tobias Siegle**, **Simon Woska**, **Jasmin Seeger**, **Tobi Renz**, **Tiko Mwakyusa**, **Markus Neuwirth**, and many more.
- All members of the **CISOVSKIT team** for the very uncomplicated atmosphere, successful scientific collaboration, and great teamwork. It was a pleasure working with you!
- **Dr. Ulrich Paetzold** and all members of the **perovskite "taskforce"** at LTI for the successful collaboration. Special thanks go to **Tobias Abzieher** and **Jonas Schwenzer** for their support in various aspects including the fast supply of samples.
- All colleagues who made the time at international conferences very special – both scientifically and personally – including **Tobias Abzieher**, **Jonas Schwenzer**, **Philipp Brenner**, **Markus Neuwirth**, **Jasmin Seeger**, **Julia Zillner**, and many more.

- **Meltem F. Aygüler** for your continuous support and the great collaboration. Thank you for many helpful discussions about perovskite chemistry, any proofreading and paper corrections, fast sample preparation, and, in particular, the fruitful time of XRD measurements at LMU.
- **Dr. Jonas Hanisch** for performing the ToF-SIMS measurements and fruitful discussions about ion migration in perovskite solar cells. **Bettina Rendenbach** for measuring the temperature-dependent X-ray diffraction patterns.
- **Meltem F. Aygüler**, **Nadja Giesbrecht**, and **Yinghong Hu** for preparation of various modifications of solar-cell samples and perovskite thin films. **Moritz Schultes** and **Ihteaz M. Hossain** for processing the semitransparent devices. Furthermore, **Tobias Abzieher**, **Jonas Schwenzer**, **Sara Moghadamzadeh**, and **Saba Gharibzadeh** for fabrication of different types of solar-cell samples.
- **Michael Hippe**, **Werner Gilde**, and **Helmuth Lay** of the electronics workshop for all competent IT and electronics support far beyond "only soldering".
- **Frank Landhäußer** and the **mechanics workshop** as well as the **joinery** for fast realization of every kind of special lab equipment and **Heinz Leonhard** for the technical drawings and lab support.
- **Our secretaries** for help and support in all kinds of administrative issues.
- The **German Federal Ministry of Education and Research** (BMBF) and the **Karlsruhe School of Optics and Photonics** (KSOP) for funding.
- All people who proofread this thesis and helped to eliminate all kinds of errors, especially **Simon Woska**, **Jasmin Seeger**, and **Lizzie Tafili**.

Additionally, support beyond the actual research project is often particularly valuable and provides necessary encouragement along the way. Beyond all already mentioned contributions in this spirit, I want to thank the following people:

- All my friends and colleagues during my studies at university for the fantastic time together, including **Alex**, **Barbara**, **Karim**, **Radius**, and many more.
- A special thanks goes to my close friends who shared a great time with me since school until today. In particular, **Dominik**, there are no words for the last years in which I had the privilege to experience a truly extraordinary friendship.
- All my **close friends spread all around the globe** from Karlsruhe and Remchingen to Florstadt, Hanau, and around, to the rest of Germany, and even Uganda, Liberia, New Zealand, and beyond. Thank you for all your support, inspiration, and prayers.
- And, of course, **my family** including **my parents**, **my sister**, and **my grandparents** for lifelong support, encouragement, and love. You're the best!

Finally, but most importantly, how could one study the influence of light without the **Father of light** himself?

“Every good and perfect gift is from above, coming down from the **Father of light**.”

– James 1:17, The Bible –

“Thanks be to **God** for his indescribable gift!”

– 2 Corinthians 9:15, The Bible –